RAND

ISM-X Evaluation and Policy Implications

Marygail K. Brauner, John R. Bondanella,
Joseph G. Bolten, Lionel A. Galway,
Ellen M. Pint, Elizabeth H. Ondaatje,
Jerry M. Sollinger

Prepared for the
United States Army

Arroyo Center

PREFACE

Logistics problems during Desert Storm, coupled with the need to operate more efficiently and economically, have caused the Army to investigate alternative logistics concepts. One concept, developed by the Army's Strategic Logistics Agency (now called Logistics Integration Agency or LIA), is integrated sustainment maintenance (ISM). In the Army, sustainment maintenance is defined as all maintenance performed above the direct-support level. It includes work done by active and reserve component general-support units, aviation intermediate maintenance units, installation directorates of logistics, Army Materiel Command (AMC) depots, and assorted contractor facilities. No single Army organization is charged with management of these diverse resources.

ISM manages the repair of unserviceable reparable components. The original ISM concept featured centralized management and decentralized execution of all maintenance above direct support, both in peace and in war. As it has evolved, ISM now features a national sustainment maintenance management structure to perform the broad function of coordinating maintenance workload among depots, contractors, and installation general-support maintenance activities. The concept also calls for a Regional Sustainment Maintenance Manager in each of several regions to make decisions about repair and to balance workload among different installations. Within a region, installations would compete to become the exclusive repair site—center of excellence (COE)—for specified reparables.

As a basis for examining policies, procedures, and practices that a regional-maintenance concept might require, the ISM concept underwent a proof-of-principle (PoP) demonstration in 1994. The PoP focused on a few reparable lines (65) within the U.S. Army Forces Command (FORSCOM). In 1995, the Army decided to expand the demonstration, increasing the number of reparable lines (212) and including organizations from other major commands. The expanded demonstration, ISM-X, lasted from July 1 through December 31, 1995. This is the final report on RAND's evaluation of the ISM-X demonstration.

RAND's evaluation is a project in the Arroyo Center's Military Logistics Program. It was sponsored by FORSCOM, U.S. Army Training and Doctrine Command (TRADOC), U.S. Army Materiel Command (AMC), and Headquarters, Department of the Army Deputy Chief of Staff for Logistics (HQDA DCSLOG). The evaluation should be of interest to DoD logisticians.

The Arroyo Center is a federally funded research and development center sponsored by the United States Army.

Army readers interested in other RAND publications listed in this report's references section should contact RAND Distribution Services: (310) 451-7002 (voice), (310) 452-6915 (fax), or e-mail (order@rand.org).

CONTENTS

FIGURES

TABLES

SUMMARY

BACKGROUND AND PURPOSE

Logistics problems during major contingency operations, coupled with the need to operate more efficiently and economically, have caused the Army to investigate different ways of providing logistics support. One concept, developed by the Army's Strategic Logistics Agency, is called integrated sustainment maintenance, or ISM. In its original form, the ISM concept centralized management and decentralized execution of all maintenance above direct support, both in peace and in war.[1] As it has evolved, ISM now is a narrower program, oriented to managing the repair of selected general-support-level (GS) reparable components. It features a national sustainment maintenance management structure to coordinate repair of selected component workload among depots, contractors, and installation general-support maintenance activities. In the future, the scope of ISM may be expanded to include other workload coordination as the program evolves. The concept also calls for a regional sustainment maintenance manager in each of several regions who can make decisions about repair and balancing of workload among different installations. For specified reparable components, installations compete to become the exclusive repair site—the Center of Excellence—within a region.

The ISM concept underwent a proof-of-principle (PoP) demonstration within one major Army command—the U.S. Army Forces Command (FORSCOM)—in 1994. The PoP focused on a relatively modest number of reparable lines (65) as a basis for examining policies, procedures, and practices that a regional-maintenance concept might require. In 1995, the Army decided to expand the demonstration to include organizations from different major commands, as well as to increase the number of reparable lines (212). This expanded demonstration, labeled ISM-X, was conducted from July 1 through December 31, 1995.

---

[1]Direct-support maintenance is usually thought of as being performed in the unit area or at the site of operation or failure of a piece of equipment; equipment is generally repaired and returned to the user. During field operations, unserviceable reparable items will not usually be held at the direct-support unit if they cannot be repaired and returned to the user within 24–36 hours; however, with the focus on cost savings/avoidance in garrison operations, this practice is not always observed. Direct-support units are trying to repair as many as possible of the "cost-driver" items (i.e., those with individual high-dollar value or with high demand). Sustainment maintenance includes work done by active and reserve general-support units, aviation intermediate maintenance, installation directorates of logistics, Army Materiel Command (AMC) depots, and assorted contractor facilities.

For a more complete discussion of Army maintenance and the flow of reparable items, see Appendix A.

RAND's Arroyo Center was asked to evaluate the ISM-X demonstration. This is the final report of that evaluation.

KEY FINDINGS

Our report focuses on the areas that need to be improved to make ISM more effective. However, we point out that this analysis would have been very difficult in the absence of ISM: ISM provides a framework (installation cost-mapping process) for standardizing costs of repair across installations and major Army commands (MACOMs), and for defining consistent sets of maintenance performance data for monitoring repair times, parts and labor costs, and other performance metrics. The following are the key findings of our evaluation:

- The most significant contribution of ISM has been to establish a process for making integrated supply, maintenance, and transportation decisions that are supported by sufficient data. Before ISM, the Army had no mechanism to view clearly its total general-support component-repair requirements and capabilities, let alone manage its workload effectively. We have cited in our analysis a variety of data on performance measures and costs, and several new management processes (selection of candidate reparables, bidding procedures for Center of Excellence [COE] selection, etc.).

- ISM-X has had many successes. Maintenance processes at the COEs improved, most components were repaired faster, and the variability of the processes declined. Furthermore, coordination of repair priorities and workloading of repair facilities across MACOMs succeeded, and the improved visibility made sustainment repair more responsive.

- As is common with new initiatives, not all goals were accomplished: Return of reparable components fell below planning figures, and by some measures (e.g., days in COE and parts cost), ISM-X did not perform as well as earlier ISM periods for selected items. These areas need closer study to identify underlying causes.

- A significant benefit of ISM is that it has caused the Army to rigorously question its repair practices. Such questioning is almost the *sine qua non* of substantial process improvement. Not only is the Army looking at how and where repairs should be made, but it is also examining the extent to which components should be repaired—and whether some repairs should be made at all. The answers to these questions will require more data, but this type of rigorous analysis can only improve Army logistics practices.

- The ISM-X demonstration showed that ISM has the potential of providing improved logistics support. The Army should proceed with implementation, because the ISM structure can lead to a more responsive and efficient logistics system. It may also avoid costs, but the extent of that cost avoidance remains to be determined and, in any event, will take some time to realize.

- Finally, we note that some additional changes (e.g., setting the retail prices equal to the repair costs or adding performance criteria to the COE selection process) will be necessary if the Army is to realize the full benefits of ISM.

HOW WELL DID ISM-X WORK?

The most positive outcome of ISM is the improved visibility into and responsiveness of the maintenance system and the critical review of Army maintenance practices. Some logistics processes improved during ISM-X, and there is potential for further performance improvements. Repairing components across major commands was accomplished without excessive difficulty—something no one was sure could occur when the demonstration began. Furthermore, it took fewer hours to fix items, and overall variability of the repair resources of time, people, and parts declined. Time spent in repair increased, but only slightly.

However, for a number of items, the repair process measures were not as good during ISM-X as prior to it. For 10 to 15 percent of the ISM-X COE lines, representing about half of the items repaired during ISM-X, repair time increased during ISM-X. These lines must be studied individually to determine why performance changed. Likewise, items with significant improvements in performance metrics—time and parts cost—should be studied to provide a source of potential solutions. Also, return of reparables fell below planned targets—an important problem for the Army as a whole and critical to the resource planning that is a primary goal of the ISM concept.[2]

Because of its short duration, ISM-X could not measure the reliability of repair or the effects of an ISM structure on readiness. As ISM is expanded Armywide, reliability of repairs and readiness need to be monitored closely.

---

[2]There were three reasons why the nonreturn of reparables was a problem:

1. When reparables are not returned, the Army may have to buy new items to replace those it would have repaired if the carcass had been turned in.
2. Resource planners procure parts and schedule labor according to projections of returns. When reparables are not returned, procured parts may go unused; thus, money was spent on parts that were unnecessary and potentially not spent on other parts that were needed.
3. In ISM, cost "savings" were projected based on "planned" returns. Because the returns were less than expected, the "savings" could not be as large as anticipated. Yet those projected savings were the basis for reducing OMA budgets for FY97.

WHAT ARE THE COSTS AND BENEFITS OF ISM?

Clearly, during the ISM-X demonstration, Operations and Maintenance Account (OMA)-funded customers at the installation level avoided costs. They used fewer man-hours and parts for repairs, and repaired items that had been purchased previously from the wholesale system. However, identifying savings to the Army as a whole over the long term is difficult for several reasons. First, the ISM-X test was only 6 months long; more long-term data are needed to make reasonable estimates. Second, as ISM is expanded, it is unlikely that the cost-avoidance effect from adding more lines and installations will be linear: The items with the greatest potential to decrease costs were chosen for ISM first; by design, the non-ISM items will yield less cost avoidance when added.

Finally, all the costs of ISM need to be accounted for, as well as all the benefits. Since the Army needs to have some measure of the costs and benefits of ISM, we suggest four actions that will improve the cost-benefit estimates:

- First, because of the variability in results, continental United States (CONUS)-wide estimates should be based on a line-by-line extrapolation rather than on a constant percentage.
- Second, price and cost factors need to reflect accurately the differences between the installation level and the wholesale level.
- Third, because economic incentives will affect how people behave at all levels, data on ISM performance should continue to be collected to ensure that desired benefits are being achieved.
- Finally, the Army should perform a multiyear analysis for workload balancing; such an analysis would take into account potential additional costs, e.g., the cost of modifying facilities, resizing the workforce, developing software, and buying hardware. A multiyear analysis of this nature may lead to an expanded set of COE site-selection criteria beyond those used in the current bidding process.

HOW DOES ISM APPLY IN CONTINGENCIES?

Support for potential contingencies was not addressed in either the ISM PoP or the ISM-X demonstration. But contingencies are central to the Army's mission, and it is important to consider ISM in that light.

The Army has an opportunity at this early stage to incorporate the benefits of ISM into the development of its warfighting support doctrine. Comparing ISM plans with current Army doctrine and practice for contingencies shows some important differences, which have both organizational and resource implications. For example, by doctrine and current practice, the Theater Army Area Command (TAACOM) manages sustainment maintenance and repair in-theater. ISM-X and ISM, as planned, do not provide for this organization; rather, ISM has either the corps or Army Materiel Command (AMC) operating the ISM regions in CONUS. These organizations currently do not have the connectivity, personnel, and equipment necessary to carry out this role when deployed during contingencies. These differences need to be resolved to ensure that ISM, doctrine, and practice evolve consistently.

SUGGESTIONS FOR IMPROVING ISM

Although ISM has accomplished some significant changes in logistics support, problems remain that, unless resolved, will keep ISM from achieving its full potential.

Financial policies and incentives surrounding logistics operations need careful review. Financial policies, although not inherently part of ISM, directly affect logistics operations. Army customers, like those of other services, find wholesale prices significantly higher than retail. For this difference to cause the installations to build up their own resupply channels and repair capacities when capacity may already exist at the wholesale level probably runs counter to good management. Thus, we recommend setting retail prices equal to repair costs. Furthermore, we suggest that the Army consider making ISM repairs reimbursable work and holding installations to their bid prices—meaning that installations can charge parts and labor only up to the bid price. To make ISM repairs reimbursable work will require changes in the financial system.

Some aspects of the management of the ISM program require improvement for optimal returns. For example, item managers at the major subordinate commands must be integrated with the ISM National Sustainment Maintenance Manager (NSMM) so that repair resources can be optimally tasked and stocks of parts can be distributed where they are needed. The NSMM must also have authority to redistribute specialized repair equipment and maintenance workload, since only the National Manager will have the visibility of maintenance capability and capacity across the Army. Additionally, the COE-selection process must be improved: Both the bidding and rebidding rules should be reviewed, and performance criteria should be added. Otherwise, the Army may not continue to receive all the potential benefits of the ISM structure.

During the ISM PoP and ISM-X demonstration, the only way to save actual dollars for the Army as a whole was to reduce procurement, because, in a short demonstration, it was not advisable to reduce either personnel or capacity. However, true savings can come only when either or both procurement and infrastructure are trimmed. The Army must make some long-term decisions about the logistics infrastructure it will need in the future and take the appropriate actions to realign or reduce unnecessary portions of that infrastructure.

Contingencies are the Army's primary reason for being. A logistics system that works efficiently and cost-effectively in peacetime is important, but the wartime functioning of that system is more important. This is an opportune time to align ISM (as planned) and Army doctrine for contingencies. Such alignment requires the allocation of resources (e.g., people, funding for equipment, information management, and communications connectivity) to the organizations that may deploy. As currently evolving, the Army risks lowering combat readiness and, possibly, creating a serious imbalance in stocks—too many parts to repair components and not enough components for replacement. However, as ISM is being fielded concurrently with other logistics initiatives, the Army should conduct a formal trade-off analysis to address a combination of training, organizational changes, stockage policy, and resource allocation.

ACKNOWLEDGMENTS

We extend our thanks to the countless Army personnel involved in ISM who hosted our site visits, painstakingly provided data, and responded to ad hoc requests for information. Without their cooperation, we could not have conducted this evaluation.

We especially appreciated the assistance of the ISM Corporate Board Chair, COL Patrick Button, Director of Maintenance, AMC, and Mr. David Campbell, Office of the G4, U.S. Army Forces Command.

At RAND, several of our colleagues—including John Dumond, John Folkeson, Rick Eden, Ken Girardini, Marc Robbins, and John Halliday—generously gave of their time and expertise to review documents and briefings.

Despite the acknowledged assistance of so many, shortcomings in the study may remain. They are the sole responsibility of the authors.

GLOSSARY

ACOM	Atlantic Command
AMC	U.S. Army Materiel Command
AMDF	Army Master Data File
ASCC	Army Service Component Command
ATCOM	Aviation and Troop Command
BDS	Battlefield Distribution System
CASCOM	Combined Arms Support Command
CDDB	Central Demand Data Base
CECOM	Communications Electronics Command
CENTCOM	Central Command
COE	Center of Excellence
CONUS	Continental United States
COSCOM	Corps Support Command
CSMS	Combined Support Maintenance Shop
CUCV	Commercial Utility Cargo Vehicle
DA	Department of the Army
DBOF	Defense Business Operations Fund
DCSLOG	Deputy Chief of Staff for Logistics
DESCOM	Depot Systems Command
DLA	Defense Logistics Agency
DLR	Depot-level reparable
DOD	Department of Defense
DOL	Directorate of Logistics
DRMO	Defense Reuse and Marketing Office
DS	Direct support
EMIS	Executive Management Information System
EUCOM	European Command
EVAC	Evacuation file in EMIS
FEDC	Field Exercise Data Collection
FLR	Field-level reparable
FORSCOM	U.S. Army Forces Command
FSC	Federal Supply Class

FTP	File Transfer Protocol
G&A	General and Administrative
GS	General support
GSU	General support unit
HEMTT	Heavy Expanded Mobility Tactical Truck
HMMWV	High-Mobility Multipurpose Wheeled Vehicle
HQDA	Headquarters, Department of the Army
IMMO	Installation Maintenance Management Office
IOC	Industrial Operations Command
IRON	Inspect and Repair Only As Necessary
ISM	Integrated Sustainment Maintenance
ISM-X	Integrated Sustainment Maintenance–Expanded
KSNG	Kansas National Guard
LAR	Logistics Assistance Representative
LOGSA	Logistics Support Activity
LSE	Logistics Support Element
LSMM	Local Sustainment Maintenance Manager
MACOM	Major Army Command
MATES	Mobilization and Training Equipment Sites
MICOM	Missile Command
MIMS	Maintenance Information Management System
MIPR	Military Interdepartmental Purchase Request
MLRS	Multiple-Launch Rocket System
MMC	Materiel Management Center
MSC	Major Subordinate Command
MWR	Morale, Welfare, and Recreation
NEOF	No Evidence of Failure
NG	National Guard
NGB	National Guard Bureau
NICP	National Inventory Control Point
NIIN	National Inventory Identification Number
NRTS	Not Repairable This Station
NSMM	National Sustainment Maintenance Manager
NSN	National Stock Number
NTC	National Training Center

OCAR	Office of the Chief, Army Reserve
OCONUS	Outside continental United States
ODS	Operation Desert Shield/Storm
OMA	Operations and Maintenance Account
OOTW	Operations other than war
OPTEMPO	Operation tempo
OSC	Objective Supply Capability
OST	Order-and-ship time
POM	Program Objective Memorandum
PoP	Proof of Principle
PPBES	Planning, Programming, Budgeting, and Execution System
PP&C	Production, Planning, and Control
QDR	Quality Deficiency Report
RC	Reserve Components
RDES	Requirements Determination and Execution System
RSMM	Regional Sustainment Maintenance Manager
RSOI	Reception, Staging, Onward movement, and Integration
RX	Reparable Exchange
SARSS	Standard Army Retail Supply System
SFDLR	Stock Funding of Depot Level Reparables
SSF	Single Stock Fund
STAMIS	Standard Army Management Information System
STANFINS	Standard Financial System
TAACOM	Theater Army Area Command
TACOM	Tank and Automotive Command
TADS	Target-Acquisition Designation Sight
TDA	Table of Distribution and Allowances
TRADOC	U.S. Army Training and Doctrine Command
TSC	Theater Support Command
TXNG	Texas National Guard
UH-1	Utility Helicopter-1 (Huey)
UH-60	Utility Helicopter-60 (Blackhawk)
USAR	United States Army Reserve

1. INTRODUCTION

BACKGROUND

Despite some impressive logistics accomplishments during Operation Desert Shield/Storm (ODS), a number of shortcomings have afflicted the Army's sustainment-maintenance system:[3]

- The joint command found it necessary to coordinate with many different Army maintenance organizations and activities. For example, both the U.S. Army Materiel Command (AMC) and theater army organizations had maintenance responsibilities for the operation.
- Maintenance was performed at AMC depots and by contractors to national inventory managers at AMC subordinate commands or to weapon systems managers, who were neither in the AMC nor in the contingency chain of command. Some of the contracts were administered by continental United States (CONUS) organizations; others were managed from within the theater.
- In addition, installation Directorates of Logistics (DOLs) performed a substantial amount of general-support maintenance.

The many players complicated the coordination with the joint command and left it without visibility of the maintenance situation at any level—national, Major Army Commands (MACOMs), or local—thus making the Army's main purpose of prosecuting contingencies much more difficult.

The ISM Concept

Army leaders recognized that such a fragmented system was unlikely to meet the needs of theater commanders in future contingencies. Therefore, the Army developed a concept called integrated sustainment maintenance (ISM). As originally conceived, all sustainment maintenance would be organized under an integrated management structure. That structure would be centrally managed by AMC and would decentralize workload to

---

[3]*Sustainment maintenance* refers to all maintenance on Army equipment above the direct-support level, conducted by Active and Reserve Component General-Support (GS) maintenance units, nondivisional aviation intermediate maintenance units, installation Directorates of Logistics (DOLs), national-level maintenance management activities and depots under Army Materiel Command (AMC), and Army contractors. Appendix A describes the Army's four-level maintenance structure and the flow of reparables through maintenance and supply.

balance resource allocation, maximize use of repair capability, improve distribution of workload, and decentralize execution of maintenance work.[4]

The current ISM concept calls for three levels of maintenance managers. At the first level, Local Sustainment Maintenance Managers (LSMMs) at each installation have responsibility for assigning work to all Army sustainment maintenance units and activities in the installation's local area, regardless of component, type, or branch. The LSMMs are responsible for program execution, but they also develop programs for reparables and training requirements for transfer to the next higher level, the Regional Sustainment Maintenance Managers (RSMMs).

The RSMMs prioritize and redirect workload among the LSMMs, tailoring the LSMM reparable programs to meet regional training and readiness requirements and reduce costs. Responsibilities at this level include developing plans for improved training of reserve maintenance units and requirements for reparables not identified by the LSMMs, as well as managing LSMM capacity or capability shortfalls by cross-leveling assets, reassigning workload, and increasing attention to ameliorating shortfalls at the next level, the national level.

At the highest level, the National Sustainment Maintenance Manager (NSMM) integrates sustainment maintenance for the Army during peacetime and contingencies. This manager coordinates wholesale requirements with the national inventory item manager and provides information for making repair-or-buy decisions for reparables. The NSMM's visibility of regional reparable programs helps item managers at the National Inventory Control Points (NICPs) review repair-or-buy decisions for extending asset utilization, reducing procurement, and maximizing savings. The NSMM is intended to have visibility of both depot and national contractor maintenance support for regional end items, and assists the NICPs by facilitating their interface with the various RSMMs to obtain regional repair support.

The initial goals of ISM included (1) meeting sustainment-maintenance requirements,[5] (2) providing visibility of sustainment-maintenance capability and capacity, (3) managing the repair of selected reparable items in support of the supply system, (4) balancing and

---

[4]We note that some of these objectives are in conflict with the concept of competition and with each other. For example, balancing resource allocation and maximizing the use of repair capability may be conflicting objectives, depending on the resources and capability. During the two tests of ISM, installations did not reduce their workforce, so it was necessary to compromise some of these objectives.

[5]As part of the budget process, Army logistics personnel estimate the amount and type of sustainment-maintenance workload to be performed during the coming year. This workload, with its associated parts and labor requirements, becomes the sustainment-maintenance requirements.

reallocating sustainment-maintenance workload to meet unprogrammed or surge requirements, and (5) achieving more cost-effective sustainment-maintenance operations by eliminating costly duplicate infrastructure and workforce, among other things. There was no attempt to prioritize these goals—all were considered important.

Business Practices

After ODS, constrained budgets led policymakers to emphasize the application of business practices, such as competition of workload, rapid redistribution of assets, just-in-time parts deliveries, and cost accountability, to Army logistics operations. Such business practices had been shown to save money in the commercial sector, and the Army hoped to achieve similar benefits in its logistics operation. The different missions facing the Army following the Cold War, coupled with declining resources, made the prospect of reduced costs especially attractive.

Business practices that promote cost-effectiveness are not, in principle, inconsistent with the ISM concept. And, as it evolved, ISM strongly emphasized "saving the Army money." However, cost-effectiveness was not the reason ISM was first conceived. Furthermore, the application of such business practices in the Army is somewhat separate from ISM, since the practices can create different economic incentives that will have to be addressed if ISM is to reach its full potential.

Testing ISM: The PoP Test

The Army conducted an ISM proof-of-principle test (PoP) in FY94, involving sustainment-maintenance activities at Forts Hood, Carson, and Riley. These three installations belong to the U.S. Army Forces Command (FORSCOM). For a portion of 65 lines selected for the test, each installation was designated as the exclusive repair center—Center of Excellence (COE)—for a specific set of lines, based on its capability, capacity, and cost to repair an item.[6] One of the important management aids for the ISM PoP was the prototype, commercially designed Executive Management Information System (EMIS). An information support system, EMIS improved the RSMM's ability to identify critical demands, monitor maintenance performance, and set and revise repair priorities.[7]

---

[6]While some effort was made to be competitive in the designation of COE, there was no recognition that after initial selection the COE became a monopoly. Later, during the ISM-X demonstration, the recompetition of workload was addressed.

[7]The drawback of any information support system is information: The information the system produces is limited by the information it receives from outside sources. Thus, the information in EMIS is only as good as the data in the systems EMIS draws upon. EMIS draws data from standard Army systems, often termed legacy systems. Unfortunately, we found that the quality of the data input to

Testing ISM: The ISM-X Demonstration

After the ISM PoP, FORSCOM decided to continue the program and to increase the number of lines in the program. At the same time, the Army decided to expand the ISM concept to two other MACOMs. The expanded ISM (ISM-X) program was a demonstration of ISM at FORSCOM, the U.S. Army Training and Doctrine Command (TRADOC), and AMC commodity commands. The three original FORSCOM installations—Forts Hood, Carson, and Riley—continued to participate and were joined by another FORSCOM installation—the National Training Center (NTC). Two TRADOC installations (Forts Sill and Bliss), AMC's subordinate commands (the Tank and Automotive Command (TACOM) and the Communications and Electronics Command (CECOM)), and the National Guard also participated. And more reparable lines were added to the original 65 lines, bringing the total number of COE lines to 212. Appendix B lists all the ISM-X lines with information on the National Inventory Identification Number (NIIN), name of the item, price, and COE.

Before ISM-X, the Army had never attempted to coordinate installation sustainment-maintenance management across MACOMs. Both TRADOC and FORSCOM were concerned that their respective items would receive different treatment at the other MACOM's installations.

Initially, the ISM-X demonstration was scheduled to run from April through September 1995. However, the demonstration did not begin until July 1995 and concluded at the end of December—just 6 months.[8] ISM-X operated similarly to the ISM PoP. LSMMs were chosen from the Director of Logistics (DOL) staffs and were augmented by contractor personnel. The RSMM remained in III Corps[9] at Fort Hood and was augmented by contractor personnel. During ISM-X, the RSMM's responsibilities expanded, because he was required to manage workload across MACOMs within the region.

As in the ISM PoP, awarding of COE status occurred at the Planning, Production, and Control (PP&C) meetings. For ISM-X, these meetings were held at Fort Hood in March and August 1995.[10] Prior to the meetings, the RSMM published a list of lines that would be up for bid. Candidate reparable components were selected according to multiple criteria (support a major weapon system, comparison of repair cost to purchase cost, minimum

---

EMIS was poor. The trend, however, was positive. As the data were being used more and in different ways, there was more emphasis on getting better-quality data input to the legacy systems.

[8]The ISM PoP began November 1, 1993 and ended July 31, 1994. Not only was the PoP longer in duration than the ISM-X demonstration, it also did not cross a fiscal-year boundary.

[9]More specifically, the RSMM was a lieutenant colonel, the Assistant Chief of Staff, Materiel, from the 13th Corps Support Command (COSCOM).

[10]The ISM-X Planning, Production, and Control meetings were numbered PP&C5 and PP&C6.

demands at multiple installations, current repair program at one or more installations, maximum washout rate). These criteria are discussed in more detail in Section 3. The primary selection priority was assigned on the basis of potential Operations and Maintenance Account (OMA) cost avoidance per hour.[11]

Each LSMM could bid on any lines his DOL or GS maintenance units were capable of repairing. The bids contained the installation's repair history (number of repairs in the prior year and washout rate), the number of labor hours, and the cost of parts to repair an item to an Inspect-and-Repair-Only-As-Necessary (IRON) standard. Even though the bids included both parts and labor, the only money that was to be transferred between installations was for the parts used in repair. The cost of labor was absorbed by the COE. During both the ISM PoP and ISM-X demonstration, normal supply-system procedures were waived so that unserviceable items could be sent directly to the COE for repair and returned to the sender, rather than returning the unserviceable carcass to the supply system and immediately issuing a serviceable replacement.

Even though there was still only one region during ISM-X, and hence no role for an NSMM to monitor repairs among different regions, an NSMM office was established primarily to coordinate national-level work[12] and establish procedures for the full implementation of ISM Armywide. Throughout ISM-X, the NSMM office consisted of a skeleton crew of personnel on loan from AMC commodity commands; no EMIS was functional at the NSMM office to provide information needed for coordination of maintenance activities.

RAND'S ROLE

After the conclusion of the ISM PoP in January 1995, the commanders of TRADOC, AMC, and FORSCOM asked RAND's Arroyo Center to conduct an independent evaluation of the ISM-X demonstration. In approaching the evaluation, both RAND and the Army acknowledged that no portion of the maintenance infrastructure exists in isolation. The quality and amount of repair at the direct-support level affect the quality and amount of repair at the general-support level, just as GS repair affects depot repair. Also, program managers for weapon systems make special arrangements for GS-level repair, which must be examined.

---

[11]"Cost avoidance will be evaluated by the difference in the repair or buy cost prior to the establishment of the COE and the repair cost as a COE item repaired at only one location." U.S. Army, *Integrated Sustainment Maintenance Expanded, Demonstration Plan,* March 1995, pp. 2–5. *OMA funds* are the installation operating funds.

[12]"National-level work is that work accomplished by the sustainment maintenance activities to meet wholesale level requirements. This process will be emphasized during the ISM-X Demonstration through increased participation by the NSMM." U.S. Army, *Integrated Sustainment,* op. cit., pp. 2–8.

Consequently, we structured our evaluation of ISM within the context not just of participating organizations or even of Army logistics, but of the Army as a whole: ISM's ability to handle both peacetime and contingency operations, its short- and long-term implications, and its consistency with logistics strategic plans for the 21st century. The focus of our evaluation differed substantially from that of the previous ISM PoP evaluation.[13] In particular, as part of the evaluation, RAND assessed the ability of ISM to ameliorate some of the sustainment-maintenance problems that occurred during ODS.

RAND's objectives were to evaluate whether ISM, as implemented, could provide centralized management and workloading of Army sustainment-maintenance activities with acceptable readiness and weapon-system availability at lower total cost than that of the previous maintenance structure, and to assess the ability of such a management structure to foster efficient use of active and reserve Army maintenance equipment, manpower, and facilities. To meet these objectives, RAND focused its evaluation on three areas: (1) the effect of this new maintenance structure and the conduct of the demonstration, (2) the rationale for an integrated sustainment structure—to include the economic incentives, and (3) long-term issues resulting from the Army's dependence on an ISM structure for sustainment maintenance.[14] This report addresses all three areas.

ORGANIZATION OF THIS REPORT

ISM represents a major change in the way a large organization does business. Such changes never occur smoothly. The ISM-X demonstration had many positive outcomes and, like any new program, areas for improvement and concern. In this report, we discuss all of these aspects.

The report addresses RAND's three focus areas as identified to the ISM Corporate Board in July 1995 and discussed in this section. Sections 2 and 3 relate to the effect of the new maintenance structure, the conduct of the demonstration, and some of the rationale for an integrated sustainment structure. Section 2 discusses how maintenance management improved under the ISM PoP and ISM-X demonstration. Section 3 describes the data we used and presents the quantitative results using COE performance measures. The next three sections discuss the economic incentives and the long-term issues. Section 4 describes the costs and benefits of ISM, and Section 5 addresses ISM in contingencies, outlining several areas that need to be addressed as ISM is implemented. Section 6 lays out some of

---

[13]Marygail Brauner, "Evaluation Plan for the Army's ISM-X Demonstration," unpublished RAND research.

[14]Marygail Brauner and John R. Bondanella, "ISM-X Evaluation: Briefing to ISM Corporate Board," unpublished RAND research.

the policy issues ISM raises and our suggestions for dealing with them. Section 7 presents our conclusions.

The report also has five appendices. Appendix A describes the Army's four-level maintenance structure and the flow of reparables through maintenance and supply, Appendix B lists the lines in the demonstration, Appendix C discusses the data sources and their limitations, Appendix D interprets the box plot, the graphical tool we used to evaluate ISM performance, and Appendix E addresses ISM and the Reserve Components.

2. HOW ISM IMPROVED MAINTENANCE MANAGEMENT

ISM unquestionably improved Army sustainment maintenance management. It has forced the Army to question its own practices rigorously. As a result of that questioning, the Army has improved management tools and practices. *Tools* include better monitoring and approaches to standardization. *Improved practices* involve a collaborative approach and a framework for making better repair decisions. Taken together, these improvements add up to a process for making integrated supply, maintenance, and transportation decisions supported by sufficient data. This process may be ISM's most significant contribution.

Later sections of this report analyze various data and suggest some improvements to ISM based on that analysis. Although the rest of the report focuses on aspects of ISM that require improvement, none of that analysis could have been done in the absence of the maintenance data collected as part of the ISM implementation. This section details some of the significant improvements that have resulted from the Army's questioning of its sustainment-maintenance practices.

ISM HAS CAUSED THE ARMY TO QUESTION ITS REPAIR PRACTICES

The testing of the ISM concept in both the PoP and ISM-X demonstration has helped stimulate questioning of the Army repair processes. Where are items being repaired? Should all circuit cards be repaired at a depot, or has technology matured sufficiently so that they can now be economically repaired at installation DOLs? The Army is even questioning which repairs should actually be done, and TACOM has found it beneficial to have a contractual arrangement with the installations to give itself another source of repair.

The Army is also looking at the depth of repair. When an item is at the DOL for repair, how carefully should it be examined for worn or broken parts, and should those parts be repaired? Should maintenance personnel look only for the discrepancy that sent the item into the DOL, or should more maintenance be performed? What are the implications for quality of repair and longevity of the components?

Even though the Army lacks data to answer many of these questions, in the process of posing them it has implemented improvements.

ONE RESULT: BETTER TOOLS

Improved Monitoring of Maintenance Programs

The ISM demonstrations developed procedures to monitor sustainment-maintenance programs, from both the requirements and the performance perspectives. These procedures could not be implemented effectively without an automated system to assist the managers in their daily functions: the variety and volume of information is just too great to be collected and analyzed manually. The developmental Executive Management Information System has been a boon to maintenance managers. In addition to providing data about maintenance jobs traditionally used by maintenance managers (such as parts costs), it broadly expands the type of information available and improves the timeliness of information so that managers can monitor supply, maintenance, and transportation activities in an integrated fashion.

For example, the RSMM has access to national supply data as well as to local data; those data aid in deciding whether to procure or repair an item. Having access to information about maintenance-shop capacity and performance helps the RSMM make decisions concerning workload allocation and awarding of bids to shops that are both technically responsive and cost-competitive.

EMIS has also increased the responsiveness of sustainment repair. With EMIS, it is relatively easy to monitor the sustainment-maintenance processes and to know which items are in repair and for how long. Information about COE performance is also available through EMIS. Theoretically, it is possible to use EMIS for differential management of critical items: The LSMM and RSMM can scrutinize repairs of items considered to be readiness drivers and encourage fast turnaround times.

Data Standardization

ISM provides a framework (installation cost-mapping process) for standardizing costs of repair across installations[15] and MACOMs, and for defining consistent sets of maintenance performance data for use in an automated system (EMIS). Since EMIS and ISM draw on current Standard Army Management Information System (STAMIS) data for the most part, their contents are only as good as the underlying STAMIS data. However, the displays and analytic tools provided by ISM and EMIS bring data discrepancies to light, thereby alerting managers to those areas in which data quality must be improved.

---

[15]Prior to developing the cost maps, the Army did not know if the cost of repairing a tank engine was the same at each installation repairing that engine. With an improved framework for calculating the actual cost of repair, the Army can identify the most cost-effective source of repair.

ANOTHER RESULT: BETTER MANAGEMENT

Collaborative Management

Traditional management in the Army is hierarchical. ISM does not bypass that hierarchy but has initiated two improvements resulting in better and faster decisionmaking through collaboration. The PP&C forum in each repair region was instituted during the early stages of ISM. At the initiation of ISM-X, an ISM-X Corporate Board was formed to provide guidance, to write policy, and to act as an arbitrator if problems arose. Both bodies have been organized so that they operate collaboratively. Both include members of the active and reserve components. Some of their deliberations result in proposals that must be approved elsewhere in the chain of command. But much of the effect of their discussions is immediate, since many discussions involve better ways of doing business that can be approved by members themselves. These two groups have already improved the management of sustainment-maintenance policies, practices, and processes.

Planning Production and Control Meetings

Planning, Production, and Control meetings are chaired by the RSMM, and the membership includes each LSMM. Others attend as invited observers.[16] The meetings are held quarterly by the RSMM, with the purpose of choosing new candidate COE lines and selecting COE repair sites through a structured bidding process. Additionally, a variety of issues, such as workload allocation, quality deficiencies, and transportation, are addressed. When possible, these issues are resolved among the participants. A particularly good feature is that the LSMMs have agreed to periodic video teleconferences (once per week, if necessary), and the RSMM holds a monthly video teleconference. We interviewed the LSMMs and found that they also confer by telephone to discuss particular situations of interest. Issues are raised to the Corporate Board for resolution if they fall outside the decision authority of the participants.

This group has initiated some significant improvements to the COE bid process (the process itself is discussed in Section 3).[17] For example, during the ISM PoP and the follow-on FORSCOM ISM program, many reparables were being sent to COE sites in worse condition than expected, but not because of the fault that caused them to fail. This poor condition stemmed from two problems: damage during shipment and stripping of serviceable parts from unserviceable reparables. This worse-than-expected condition drove the cost of repair

---

[16]The RAND research team attended several of the PP&C meetings.

[17]Until the PP&Cs were implemented there was no group that could solve problems across Army installations.

beyond the bid price. One solution was for FORSCOM to buy the necessary packaging and crating equipment and materials for the installations. A second solution, implemented at no cost, was to have the sending installations implement policies to ship COE items intact, with all their appropriate parts, and to have items stored in good condition until they could be sent to a COE. (We had observed damage in earlier phases of ISM, such as vehicle engines in shipping containers in several inches of water and mud because they had been left uncovered outside before shipping.)

Another benefit is the sharing of process-improvement information. For example, one LSMM briefed the others on how he could contract for "less-than-truckload" deliveries for less than the normal cost by committing to routine shipments; he found that this practice was cheaper than premium transportation services (such as Federal Express) and still met acceptable delivery timelines.

ISM-X Corporate Board

The ISM-X Corporate Board has representatives from each major command participating in ISM-X. In most cases, the members are the principal policy developers from their major command. The board is chaired by the Director of Maintenance, HQ, AMC. Other members represent both supply and maintenance policy offices within their commands. The Army Headquarters Deputy Chief of Staff for Logistics (DCSLOG) also has representatives on the board from the supply and maintenance policy offices. Consequently, this board can resolve problems and formulate new policies or implement changes quickly.

The Chiefs of Supply from several MACOMs also sit on the board, a fact that emphasizes that ISM is not just about maintenance but is a program that can effectively integrate supply, maintenance, and transportation decisions.

Improved Focus on Efficiency

Because the awarding of COE status is based on man-hours, cost of repair parts, washout rates, and cost of transportation and packing,[18] the DOLs are eager to reduce their costs and to improve their efficiency. At every installation we visited during ISM-X, we heard discussion about how it needed to become more competitive so that it could at least maintain its workload. At the PP&C meetings, it has been clear that the LSMMs are examining their processes, capabilities, and equipment so that they can bring more work to their installations.

_____

[18]Prior to PP&C6, COE status was awarded only on the basis of man-hours and cost of repair parts.

Improved Repair-Program Decisions

The early focus of ISM has been to save operating costs at the local installation by repairing more there and buying less from the wholesale system: ISM has provided a better method for making these decisions. Instead of repairing something at one installation for that installation's own benefit, ISM can look across the Army to determine whether it is better to repair or procure an item. Thus, the local repair programs can better serve the Army as a whole. We discuss in Sections 3 and 4 some methods to improve these decisions.

Repair programs are being instituted or continued for the additional reasons of the national supply position, a reparable's contributions to equipment availability and unit readiness, and both the cost and time to repair compared with those to procure. Further, installations are changing processes to establish more-efficient repair programs within the resources they already have.

3. MEASURING ISM-X PERFORMANCE

One of the goals of both the ISM-X demonstration and the ISM concept is to improve maintenance processes through competition and specialization when work is concentrated at COEs. In this section, we use data collected during the ISM-X demonstration to analyze the maintenance process for three performance measures—days in repair at the COE, man-hours, and cost of parts—by comparing the three during the ISM-X demonstration with a baseline for the repair of the same items prior to ISM-X. The data set is described in detail in Appendix C. We also briefly discuss the distribution of transportation times during ISM-X. We then use the maintenance data to examine some important issues about another vital part of ISM: the bids made by the COEs and the return of unserviceable carcasses for repair.

Because quality of repair and the effect of maintenance on readiness are two issues of enduring concern to the Army, it is important to measure how the ISM structure affects both. However, these two issues are difficult to measure in a six-month demonstration; therefore, we describe how the Army can collect information for monitoring both quality of repair and readiness. These discussions naturally lead to the topic of data quality and its importance to the future evolution of ISM. This section concludes with a list of issues and suggestions to be addressed as ISM is implemented.

COE PERFORMANCE MEASURES: COE DAYS, MAN-HOURS, PARTS COST

The three primary measures of repair-process performance are total days in the COE, man-hours, and parts cost. The primary goal is to reduce the typical, or average, value for all three for each line in ISM-X while maintaining quality of repair.[19] A secondary goal is to reduce the variability in the measures for each line.[20] We examine both of these goals below.

---

[19]The ideal measure of ISM-X performance is overall repair-cycle time (Marc Robbins, "The Need to Measure Repair Cycle Time," unpublished RAND research). *Overall repair-cycle time* is the time from diagnosing the item as "broken" until it is serviceable and ready for issue (or, under the policies used during the ISM-X demonstration, returned to the user). This includes time spent at the division and below in diagnosis and item movement, transportation time to the COE, days at the COE, and movement back to the unit or to supply. Unfortunately, as noted in Robbins, it is very difficult to get reliable information on the lengths of some of these segments. In any case, the measures we have selected are of interest in themselves as indicators of the individual pieces of the repair pipeline.

[20]Current quality practices (see, e.g., Howard Gitlow, Shelly Gitlow, Alan Oppenheim, and Rosa Oppenheim, *Tools and Methods for the Improvement of Quality*, Richard C. Irwin, Homewood, IL, 1989) emphasize both goals because customers usually want repair times to be both short and predictable (have little variability). The latter goal is secondary in repair because, while desirable, the inherent variability in failure modes suggests that repair time also has some inherent variability.

- 14 -

Our decision to assess maintenance performance in ISM-X by looking at the distribution of measures for each line requires some comment, because the Army's approach, while easy to understand, does not provide enough information for decisionmaking. The current approach to assessing performance in ISM-X is to compute the mean for each of the three measures (especially days in the COE) across all ISM lines repaired during the demonstration period and to compare them to the average for the same measure during some previous period. This approach *does* produce a single number as an overall measure of performance; however, the procedure suffers from the disadvantage that the two means are not directly comparable unless the mix of items repaired is the same during both the test and comparison periods. It is not difficult to construct hypothetical examples in which, by changing the mix of items in the two periods, improved process performance can show a worsening of the average measure and degraded performance will show an improvement.

Typical Performance

We assess the *typical* performance of the repair process during the ISM-X demonstration period by computing the median[21] for each measure for all repairs on a given COE line at the COE during the demonstration period. The baseline, or comparison, value is the median of the corresponding measure for all repairs done on the same COE line at all installations participating in ISM-X in the "historical" period (9 months) before the start of ISM-X. Since the performance measures vary widely for the diverse lines in ISM-X, we have computed a standardized variable by dividing the median performance during ISM-X by the historical median. A value of one indicates that the performance of the COE during ISM-X is the same as that prior to ISM-X. Values of less than one imply improvement; values greater than one imply worse performance. For example, the M939 starter engine (NIIN 00-304-3493), which was assigned to the COE at Fort Bliss at the start of the ISM-X demonstration, had historical median man-hours to repair of 2.8 (i.e., half the repairs took 2.8 man-hours or

---

[21]In this analysis we use the median (50th percentile) of the measures for each job because, unlike the more familiar mean or arithmetic average, the median is a more stable measure of the center, or typical, value, loosely speaking, of the distribution. By "stable," we mean that very large values of time, man-hours, and parts costs (which are likely to be data errors) do not influence the median as heavily as they do the mean. Such large values should not be ignored in more detailed analyses of the data, but we do not want them to obscure the overall typical performance of the repair process. For the same reasons we later use the interquartile range (the range between the 25th and 75th percentiles) as a measure of variability instead of the more common standard deviation. As with the median versus the mean, the interquartile range is less vulnerable than the standard deviation to data problems. Note that the Velocity Management initiative also uses percentiles to measure the performance of various logistics process measures. For more details on the mean and interquartile range, see Paul F. Velleman and David C. Hoaglin, *Applications, Basics, and Computing of Exploratory Data Analysis,* Boston: Duxbury Press, 1981.

less). During ISM-X, the median declined to 2.5 to give a ratio of 0.89, indicating an overall improvement in man-hours to repair.

In our original analyses,[22] we used only completed jobs to determine the shift in COE hours, man-hours, and parts costs. However, as we noted at that time, a more sophisticated analysis would take into account the open work orders as well, since these incomplete jobs contribute information that the job will take at least as long as the time the work order has been open. (Because the ISM-X demonstration period was so short, using all available data is important for evaluation.)

We used nonparametric statistical techniques from survival analysis to compute the median days in the COE during the ISM-X demonstration, for each COE line, using both the completed and uncompleted jobs.[23] However, we used only completed jobs to determine the *historical* median, because jobs that began before the start of the demonstration and were still open as of six months after the start of ISM-X were highly likely to be data errors or have other special characteristics. We also used only completed jobs for man-hours and parts costs, because we had anecdotal information of varying policies among different installations for entering these data elements in records of open jobs. Examination of our data, which showed that most of the incomplete jobs had zero man-hours and parts but some did not, supported this approach.

To ensure that our estimates of the two medians for each COE line were based on sufficient data, we restricted our analyses of these performance measures to COE lines that had at least 10 completed repairs both before and after the start of ISM-X. In our analysis of man-hours and parts, we eliminated those jobs with zero man-hours and parts, because we were unable to determine whether those data were missing or were really zero (zero parts cost is possible during the repair of some types of items). For this reason, the number of COE lines used may vary among the different performance measures and is certainly far less than the total number of lines in the ISM-X demonstration.[24]

Figure 3.1 displays performance measures for the new COE lines—those lines that were awarded to COEs for the first time during the ISM-X demonstration. For each of the

[22]Marygail Brauner et al., "ISM-X Preliminary Findings: Briefing to ISM Corporate Board" unpublished RAND research.

[22]Marygail Brauner et al., "ISM-X Evaluation and Policy Implications for ISM Implementation," unpublished RAND research.

[23]Rupert G. Miller, *Survival Analysis*, Wiley, NY, 1981. Miller discusses how incomplete, or censored, observations contribute to the determination of length.

[24]The reader will want to note the number of items included in each analysis. For example, of the 91 new lines in ISM-X, only 22 in our data had 10 or more completed repairs, both before and after the start of ISM-X, and nonzero man-hours.

three performance measures, a boxplot indicates the median, 10th percentile, and 90th percentile of the distribution of the median ratios.[25] Below the boxplot are the number of COE lines represented in the plot; these are the new COE lines that met our criteria for analysis.

In Figure 3.1, the boxplots for both man-hours and parts costs show that the median of the ratios is less than one; for man-hours, 75 percent of the lines were repaired with fewer man-hours. However, the overall calendar time spent in repair increased slightly over history. As indicated by the outlying dots, the variability in the ratios—from near zero to over 40—is significant. While many lines were real success stories in that their repair performance improved dramatically, others showed worse performance.

Figure 3.2 shows the same information on days in the COE, man-hours, and parts costs for the COE lines continuing in repair during ISM-X that met our criteria for analysis (these were lines from the ISM PoP and from intervening PP&C conferences). Here, we

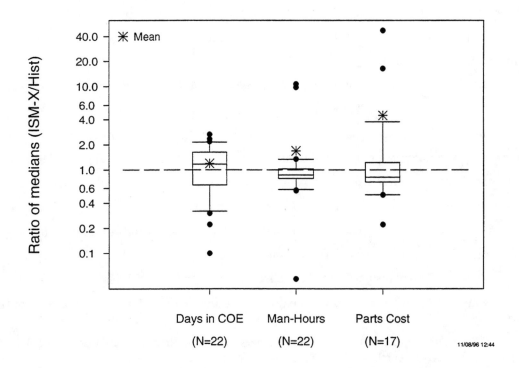

Figure 3.1—Performance Measures for New Lines, ISM-X Versus History

---

[25]The asterisk indicates the mean, and the dots represent repairs in the bottom or top 10 percent. Appendix D graphically explains a boxplot and the information it displays.

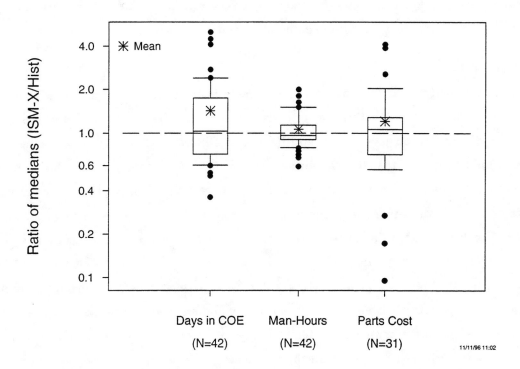

Figure 3.2—Performance Measures for Continuing Lines, ISM-X Versus History

might expect less overall improvement and closer clustering around one, since these lines had already been repaired at COEs and some process improvements had already been made. As expected, the medians of the distributions for each of the three measures are near one, although, again, there is substantial variability.

Both Figures 3.1 and 3.2 use a logarithmic scale on the *y*-axis. (The reason for a logarithmic scale is explained in Appendix D.) For the new COE lines shown in Figure 3.1, the ratio of parts costs has two significant outliers, meaning that for two of the new COE lines the parts costs were far greater than the parts costs during the pre–ISM-X period. Man-hours also has two significant outliers. For the continuing COE lines shown in Figure 3.2, it is notable that three had markedly lower parts costs than in the previous period.

A number of factors could potentially affect each of these measures beyond the maintenance practices of the COEs, including changes in level of repair needed, as a result of changes in use; differences between repair practices at the designated COE and the other installations that repaired an item in the baseline data set; lack of available parts for some items (particularly more-complex items in aviation); and the disruptions caused in late 1995 by the partial government shutdown during the budget-approval process. What needs to be done now is to look at the lines whose repair improved, noting the factors that made the improvement possible; focus on the lines whose repair processes did not improve; and analyze the reasons for the increased time, man-hours, and/or parts costs.

Variability

A second way to judge process improvement is to look at the variability of the processes, i.e., how widely the number of days at the COE, man-hours, and parts costs vary over individual repair jobs for a given ISM-X line. Different repair jobs on the same type of item can be expected to take different amounts of effort, parts, and time because items can fail in different ways. But the COEs may be able to eliminate some of that variability for the items by improving diagnostic skills through specialization and experience, by acquiring better tools and instituting better parts stockage, and by focusing repair resources to keep a repair job from being interrupted.

As with the improvement metrics above, we constructed a standardized variability-improvement metric between ISM-X performance and historical performance by taking a ratio between variability measures for each line for the two periods. For the variability measure in each period, we used the difference between the 75th and 25th percentiles[26] (the "interquartile range") for each performance measure before and after the start of ISM-X. Figure 3.3 displays boxplots of the distribution of the ratios for all ISM-X lines (both new and continuing) for days in COE, man-hours, and parts costs. Over 50 percent of the lines showed a decrease in the spread for each measure (ratio less than one), indicating a reduction of variability in the repair process. And for man-hours, 75 percent of the lines showed a decrease in the variability. As with the median improvement measures, however, substantial variability occurs across lines.

On balance, the COEs have managed to make the repair process less variable for a majority of items. Removing all variability from the repair process is not likely, but the challenge for the COEs and ISM will be to glean lessons from the items that have improved and to strengthen the options they have for overcoming variability in repair (e.g., rapid ordering of parts, collocating repair facilities that can repair multiple lines).

---

[26]The 25th percentile of a set of data is defined as the value for which 25 percent of the data lies below the value. The 75th percentile is defined analogously. (As noted above, the median of the data is the 50th percentile.) The difference between the 75th and 25th percentiles tells how spread out the central 50 percent of the data is (somewhat like the variance, but this difference of percentiles is, like the median, more robust to bad data).

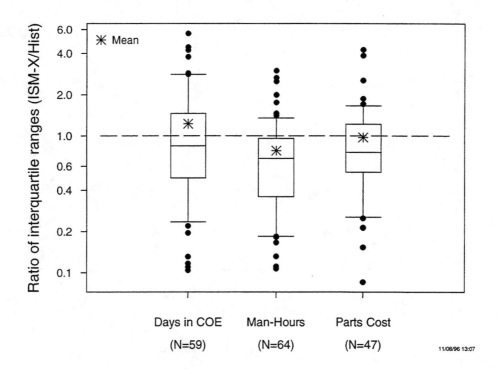

Figure 3.3—Variability of Processes for All Lines, ISM-X Versus History

Price as an Explanation for Variations in Performance Measures

We briefly explored the possibility that item price (as a rough surrogate for item complexity) might give some indication of why the repair of some lines improved during ISM-X while others did not. Figures 3.4, 3.5, and 3.6[27] show the median ratios for COE days, man-hours, and parts costs plotted against the Army Master Data File (AMDF) price for each line.[28] Somewhat surprisingly, the plots show no strong relationship to price. In fact, some of the least-expensive lines (those that plot in the upper left-hand corner of the box) had the highest ratios of ISM-X to historical performance.

---

[27]As with the previous figures, we have plotted the ratios on a log scale to fit all the data within the plot and to give equal visual weight to items that improved. We have also plotted the Army Master Data File (AMDF) price on a log scale because of the wide range of prices for items being repaired during ISM-X.

[28]The AMDF price was taken from the December 1995 version of the AMDF. When the prices of substitutables varied for a line, we took the highest price.

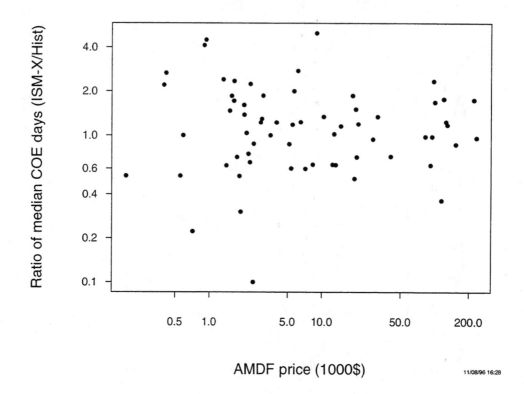

Figure 3.4—Days in COE Versus AMDF Price

We did not attempt to pursue further the causes behind the increase in performance measures, primarily because to do this kind of detailed analysis would require focusing on each line and understanding in detail its repair process, parts and skill requirements, etc. This is a job that the RSMM and LSMMs are doing as part of the ongoing management of ISM; these plots could be a useful tool in helping the RSMM and LSMMs identify these lines.

Days in COE as a Performance Measure

Days in COE is an important performance measure because it directly measures the delay in converting an unserviceable item into a serviceable one (Figure 3.7). Note that actual man-hours being used to repair an item is a very small fraction (1–10 percent) of the time the item is at the COE. Said another way, 90–99 percent of the time an item spends in the COE, no man-hours are being expended on it. Also as shown in Figure 3.7, there is little correlation between man-hours in repair and the number of days an item spends at a COE. There is room for improvement in overall repair times.[29]

---

[29]As with the "canonical" ISM-X performance measures, we did not investigate the specific problems pertaining to each part. These could include administrative delays, time spent awaiting parts, etc. These pipeline times are being addressed by the RSMM and LSMMs.

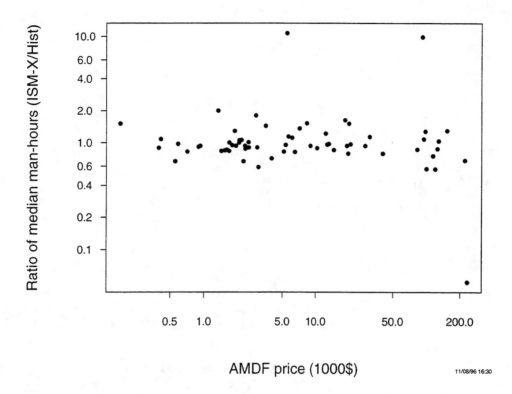

Figure 3.5—Man-Hours Versus AMDF Price

In the figure, we have plotted the median calendar time in the COE during ISM-X for the lines meeting our criteria against the median man-hours used for their repair (as a rough surrogate for complexity of repair). It is clear that some parts, e.g., the M35 engine or the AH-64's Target Acquisition/Designator System (TADS) turret, take considerably longer to clear the COE than would be expected from the man-hours required to complete their repair. Again, the lines that stand out are candidates for close attention to determine and fix any obstacles to repair.[30] To improve performance, the Army might consider using "Days in the COE" as a bid parameter. By doing so, the ISM program would encourage LSMMs and RSMMs to determine the causes of these unexpected delays and improve the underlying process.

---

[30]Boxplots showing the performance measures for all jobs for each item are contained in a supporting document to this report: Lionel A. Galway, "ISM-X Evaluation: Individual Item Repair Performance Measures," unpublished RAND research.

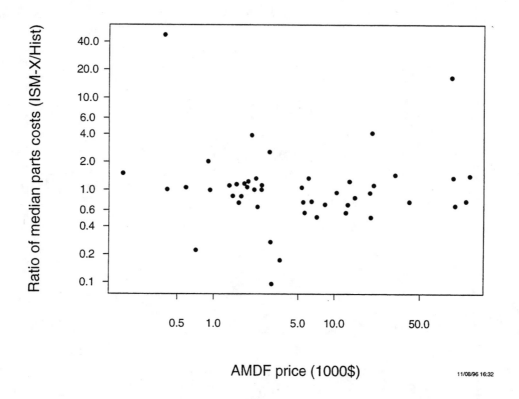

Figure 3.6—Parts Costs Versus AMDF Price

TRANSPORTATION TIMES

Figure 3.8 shows the distribution of transportation times during ISM-X for shipment from the installation to the COE, and from the COE to receipt by the installation (data are from the EVAC file—one of the files in EMIS; see Appendix C for more details on data), for each of the valid shipments.[31] Ninety percent of the shipments in each direction were completed in well under 20 days, with the median shipment being completed in under 10 days. In discussions at the PP&C conferences we attended, the performance was considered

---

[31]This figure eliminates records that had zero entries in either of the two endpoints, as well as negative times and times over 80 days (4 records met the latter criterion for exclusion). Records for in-transit and in-repair parts will be necessarily incomplete; however, records with missing dates in some segments followed by segments with dates clearly have problems. We have not quantified the problems here, except to note that the segments from the installation to the COE have better data, particularly for the two segments we chose. Since much of this data entry was manual during ISM-X, there are opportunities for mistakes and omitted entries.

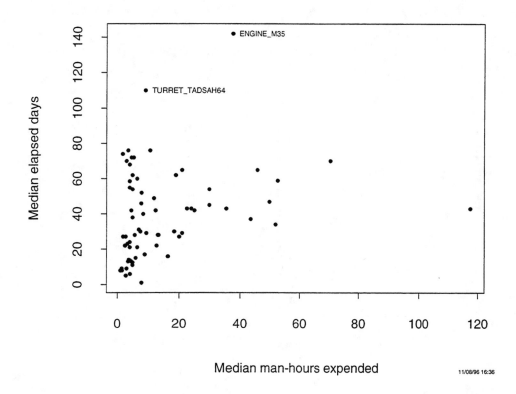

Figure 3.7—Median Days in COE Versus Median Man-Hours Expended

good, although it is somewhat difficult to define a baseline against which to compare these times.[32] A recent study of the Army's order-and-ship-time (OST) process[33] found a median time of 19 days for OST (from order to receipt) of high-priority serviceable lines, with a 75th percentile of 36 days. The ISM-X transportation processes are better both in median and variability of transportation times.

We note that it was not possible to examine transportation costs for ISM-X, because these data were not collected in a form that allowed us to separate them from transportation of other items to and from an installation. The cost of transportation is an important metric that the Army must track if the savings and benefits of ISM are to be correctly measured.

---

[32]During ISM-X, shipments were made using Federal Express, when possible—in effect, standardizing decisions about transportation mode. In addition, handling was expedited at both the originating and receiving installation.

[33]Kenneth Girardini et al., "Measuring Order and Ship Time for Requisitions Filled by Wholesale Supply," unpublished RAND research.

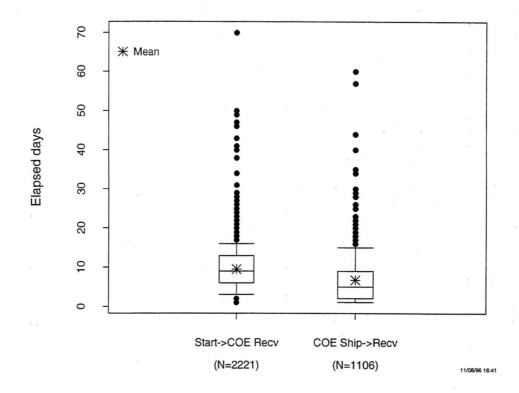

Figure 3.8—Transportation Times to and from COE

ACTUAL RETURNS OF UNSERVICEABLE CARCASSES FELL BELOW PLANNED RETURNS

During the ISM-X demonstration, actual returns of unserviceable carcasses fell below planned returns.[34] This is a potential problem, because the planned returns were used as the

---

[34]Several different research efforts have documented the large variability in removal rates of components in military equipment (see, e.g., Marc Robbins, *Developing Robust Support Structures for High Technology Support Systems: The AH-64 Apache Helicopter*, Santa Monica, CA: RAND, R-3768-A, 1991; James Hodges, *Onward Through the Fog: Uncertainty and Management Adaptation in Systems Analysis and Design*, Santa Monica, CA: RAND, R-3760-AF/A/OSD, 1990; and Gordon Crawford, *Variability in the Demands for Aircraft Spare Parts*, Santa Monica, CA: RAND, R-3318-AF, 1988). However, the fact that items with larger demands are uniformly underestimated suggests a systematic bias that needs to be investigated. Shortfall in returned unserviceables is a serious issue for all Army repair activities both in peace and in conflict, because such a shortfall limits repairs due to lack of unserviceables. It has also been cited as a problem by the other services. For a discussion of this issue during Desert Shield/Storm, see James A. Winnefeld et al., *A League of Airmen: U.S. Air Power in the Gulf War*, Santa Monica, CA: RAND, MR-343-AF, 1994; and several U.S. Army after-action reports, among them *22d Support Command After Action Report*, Volume 5, ODS 838681, 1991. For a discussion of this problem in the Navy, see Lionel Galway, *Management Adaptations in Jet Engine Repair at a Naval Aviation Depot in Support of Operation Desert Shield/Storm*, Santa Monica, CA: RAND, N-3436-A/USN, 1992.

basis of workload planning, parts buys, and estimation of anticipated ISM-X cost avoidance. In Figure 3.9, we display data about the returns during the ISM-X demonstration. We divided the actual returns during ISM-X by the planned returns.[35] A ratio of 1.0 (indicated on the plot by the straight center dotted line) means that the number of unserviceables for a line exactly matched the plan; a ratio greater than 1.0 indicates that more unserviceables were returned than planned; a ratio less than 1.0 indicates a shortfall of unserviceables. We plotted this ratio against the number of the planned returns. (In general, it is easier to have planned and actual numbers close together when many returns are planned and more difficult as the planned numbers approach 1 and 2.) In Figure 3.9, each dot represents an ISM-X COE line. The curved upper and lower dotted lines are a rough 95 percent confidence limit for the actual returns.[36]

Figure 3.9 illustrates that while a number of lines met their targets, a substantial number did not. Some lines were, in fact, returned in greater numbers than planned; however, all but one of those lines were fairly low-rate. The most important feature of the graph is that, for more than 40 planned returns, all but one of the lines that missed its target fell below by fairly substantial amounts.[37]

There could be many reasons for the shortfalls, ranging from low numbers of failures because of seasonal training patterns, to problems with the federal budget, to failure of units to turn in unserviceables. The causes of the shortfalls need to be investigated, particularly if budgets are set according to planned numbers.[38]

The general topic of return of unserviceables is discussed in Marc Robbins, "Improving the Army's Repair Process: Baseline Repair Cycle Time Measures," unpublished RAND research, and in Irving Cohen, *Coupling Logistics to Operations to Meet Uncertainty and the Threat (CLOUT): An Overview*, Santa Monica, CA: RAND, R-3979, 1991.

[35]The planned number of repairs was computed by prorating the annual planned workload over the months of the ISM-X demonstration, which is not quite correct: The planned returns for some items were not expected to be uniform over the year. But this approximation should be good enough to demonstrate the overall trend.

[36]These confidence limits were computed by assuming that the returns were generated by a Poisson process operating at the planned rate, which assumes independence of demands. This assumption is not quite correct but gives a rough indication of how much variability to expect.

[37]The exceptional point is the High-Mobility Multipurpose Wheeled Vehicle (HMMWV) fuel pump repaired at Fort Hood, for which the planned returns were 140. The number repaired was almost three times what was planned. We queried the LSMM about whether the pump was experiencing some special problem, but the best assessment they had was only that a very high number had been returned.

[38]We note in Section 4 that calculation of ISM "savings" was based on planned returns. The actual returns fell far below those planned, which implies that actual savings would also be less than planned savings. This is a serious problem for installations if their budgets are reduced prior to achieving the savings.

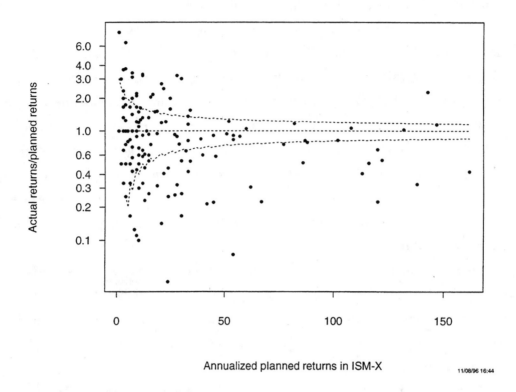

Figure 3.9—ISM-X Planned and Actual Repairs for All ISM-X Lines

INTER-MACOM AND INTERINSTALLATION SUPPORT

A concern during the planning of ISM-X was whether inter-MACOM repair would be as responsive to the user as intra-MACOM repair. Unfortunately, very few ISM-X lines were repaired across MACOMs in sufficient quantity to provide reliable estimates of the differences in performance among intra-MACOM, inter-MACOM, and the own-installation jobs. (We required a minimum of 10 jobs in each category to do meaningful comparisons.) Informally examining individual lines showed no pattern of differences.

We also examined a rough surrogate measure for inter-MACOM support: During the ISM-X demonstration, 19 ISM-X COE lines had 10 or more repairs from both the COE installation and other installations. For each line, we compared the median time spent in the COE for on- and off-installation repairs (regardless of MACOM): 74 percent of the lines had the same or better median times for off-installation repairs; 26 percent had worse times for repairs done for other installations, indicating that, on the whole, the COEs supported other installations as well as their own.

QUALITY OF REPAIR AND READINESS

Quality of repair under ISM-X and the effects of ISM-X on the readiness of supported units were serious issues that were much discussed before the beginning of the ISM-X demonstration. Ideally, we had hoped that the demonstration would provide insights into how these two issues would fare under the ISM-X concept. However, we did not evaluate either of these issues during the demonstration.

For quality of repair, the period of the demonstration was simply too short to collect enough data to detect, if present, repeated returns of a supposedly repaired item. Furthermore, serial-number information on repaired items (which is needed to identify a repaired item over several different repair jobs) is not collected consistently by Army data systems. This latter point also means that it is difficult to assemble baseline data on the rate of return of items considered repaired under the pre–ISM-X maintenance structure.[39]

We also did not evaluate unit or equipment readiness (for equipment with components in the ISM-X demonstration) because of the short time for the demonstration. Because most units have a yearly cycle of activity with periods of greater and lesser activity, we believe that a minimum of one year is required to detect all but the most severe effects on readiness at the system level. (In addition, the ISM-X demonstration period included the end of the 1995 fiscal year and, as noted before, the protracted struggle over the 1996 budget, with continuing resolutions and the partial government shutdown.) Furthermore, resourceful units can keep equipment-readiness rates high for several months by skillfully using their available resources, e.g., having maintenance personnel work overtime, borrowing spares from other units.[40]

Both issues will need to be monitored carefully as ISM is implemented and a longer series of data becomes available for analysis.

CHARACTERISTICS OF COE BIDS IN ISM-X

Bidding for COE work is a key component of ISM-X (and future expansions of the ISM concept, as now conceived). Installations bid for new workloads based on their past experience and on potential improvements in the repair process. However, in examining the detailed individual data presented above, we noted a number of cases in which the bids for man-hours and parts costs differed greatly from the quantities recorded both for repairs in

---

[39]Our understanding is that the RSMM was tracking QDRs (Quality Deficiency Reports) for ISM-X items as a method of tracking repair quality.

[40]During the ISM PoP, the overall readiness rate of supported equipment was monitored, and, according to Army documents, little effect was found.

the baseline data and for those actually performed by the COE on the item during ISM-X. This section discusses bids with respect to both history and ISM-X performance.[41]

As with some of the previous performance measures, we used ratios of median historical performance for all installations participating in ISM-X (prior to the ISM-X period) and median ISM-X performance of the winning bid made by the successful COE. These ratios put the information for all COE lines onto a single scale. Figure 3.10 plots the ratios of the bid for man-hours and parts costs to median historical performance, restricted to COE lines with at least 10 repairs in the nine months preceding the start of ISM-X. In this plot, points above the dotted line at 1.0 indicate those winning bids for repairing COE lines that were higher than the historical median; points below indicate COE lines for which the winning bids were lower. While the median ratio is almost exactly 1.0 for each measure, a great deal of variability becomes apparent in the bids when they are viewed in historical

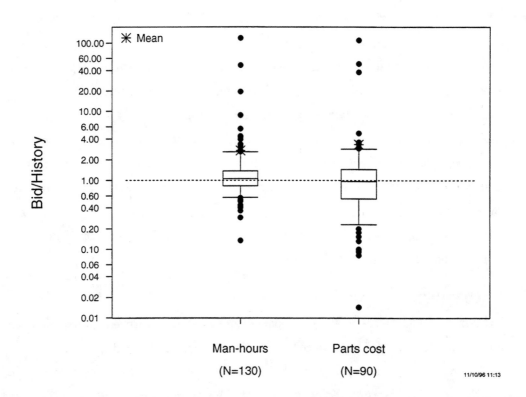

Figure 3.10—ISM-X Bids Versus History

---

[41]Detailed information on the bids for each individual COE line and their relation to historical and actual performance is available in Lionel A. Galway, "ISM-X Evaluation: Individual Item Repair Performance Measures," unpublished RAND research.

perspective. Note that the goal should be to get as many of these ratios as possible to be less than or equal to 1 (i.e., the COEs' goal should be to match or improve on historical performance).

Figure 3.11 plots the ratios for ISM-X median performance divided by bid for each COE line having 10 or more repairs during ISM-X. Here, points below the 1.0 line indicate COE lines for which the COEs did better than their bid; points above the line indicate COE lines for which the median performance was worse. The medians of the distributions are improved, but, once again, there is significant variability, with almost 25 percent of the lines having median parts costs exceeding the acceptable limit of 125 percent of the original bid.[42]

Clearly there is much more detail here than we could explore. For example, the bids and performance could be examined by COE or by item cost to see if bid problems are characteristic of certain COEs or of particular lines (or both). In any case, the wide

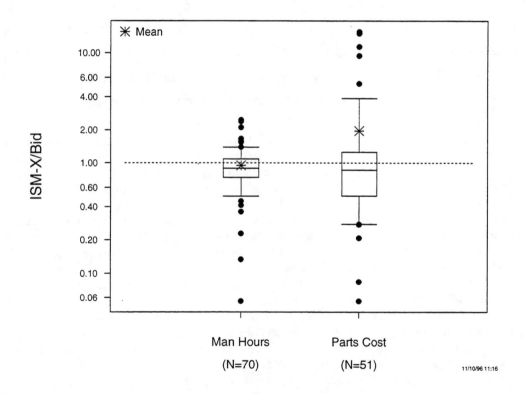

Figure 3.11—ISM-X Performance Versus Bids

---

[42]During ISM-X, an item was not removed from a COE and rebid as long as the average performance of the COE was under 125 percent of the bid for both man-hours and parts costs.

variability is cause for concern about how the bids are formulated and how COEs are meeting the challenges. Graphics like these could be used by the RSMM and LSMMs as a screening device for focusing attention on potential problems during the bid-submission process.

As noted earlier in this section (Figures 3.1 and 3.2), actual man-hours during ISM-X were marginally better than historical man-hours, but there was considerable variability in actual-versus-historical man-hours. This variability appears to cause bidders to err on the side of caution as shown in Figure 3.11, where actual performance is better than the bid.

DATA QUALITY

The analyses presented in this section are conditional on the quality of the data collected by EMIS from the maintenance legacy systems and the various manually inputted data. In the course of our analysis with ISM-X data, we encountered a number of data-quality problems. Most were fairly minor (e.g., typos, inconsistent quantities for the various disposition categories such as repaired, Not Reparable This Station (NRTS) and washout), but some were potentially more serious: For example, we found that some closed work orders on repaired ISM-X parts at COEs had zero parts costs[43],[44] and that there were missing or out-of-sequence dates in about 20 percent of the records in the EVAC file.[45]

---

[43]During the ISM-X demonstration, the only money changing hands between installations was for parts, which should have created strong incentives to record parts costs accurately.

[44]This latter point has caused extended discussion; therefore, we give a few more details. In our examination of COE performance, we considered a job as "closed" if it had either a "finish work" date (preferred) or a "pickup" date in the WORKORDERS file. We noted, however, that a substantial number of such jobs had a zero parts cost. An examination of the data (through January 2, 1996) for ISM-X items (at COEs and non-COEs) showed that about 20 percent of these jobs had zero parts cost. Restricting the data analyzed to the ISM-X demonstration period (jobs opened after July 1, 1995) reduced the percentage to 17 percent. Further discussion with the RSMM indicated that we should only consider jobs as closed if they had an "S" status, and that we should restrict our attention to only COE work. (Work on COE items at non-COE sites during ISM-X were probably predominantly inspections or simple adjustments. For example, one work order at a non-COE site was for a job whose activity was palletizing for turn-in COE components from a deactivated unit.) With this restriction, the percentage dropped to 13 percent. The percentage of missing values also varies by materiel category code (electronics parts have a high percentage of zero parts costs, which is consistent with the contention that many "repairs" are actually adjustments). Conversations with RSMM personnel indicated that they have flagged some COE jobs with zero parts, called the LSMM and received a corrected number, which has been entered in some of the specialized databases (e.g., cost avoidance) maintained by the RSMM. (Currently, the WORKORDERS file in EMIS is not corrected by the RSMM, because it is overwritten nightly with a new file from the LSMM EMIS systems.) On the other hand, some jobs did have legitimate zero parts costs.

[45]Note that much of the data entry into the EVAC file for shipping-related dates was manual during ISM-X.

The range of problems and the fact that our analysis utilized only a few selected data elements from the WORKORDERS and EVAC files lead us to suggest that the importance of data quality needs to be stressed as ISM proceeds to be implemented.[46] As long as decisions are being informed by data, the RSMM must continue to be aggressive about enforcing data quality and encouraging corrections. The data used for analysis, comparison of the various repair activities, charges to customers (such as labor hours, parts costs, and transportation), and cost projections must reflect reality; otherwise, the analysis will be suspect, the potential for incorrect decisions will exist, the cost of repairs will not be appropriately allocated, and budgets may not be sufficient for the required mission.

ISM-X is hardly unique for having problems with logistics data quality that have large effects on management decisions.[47] EMIS, in its role as "front end" for a variety of legacy systems, is limited in what it can do to correct problem data from the COEs. Indeed, the use of EMIS by the RSMM to flag such problems indicates its value in helping to improve data quality.[48]

While our use of medians (and our exclusion of zero-parts-cost jobs) protects us to a certain extent from having our various performance measures biased by such jobs, this caution illustrates the problem of using data from the legacy systems for such comparisons. And yet, without being able to use historical and current data for comparisons, it is hard to tell if processes are improving.

SUGGESTIONS FOR IMPROVING ISM

Below, we summarize some points suggested by the analysis in this section that should be addressed as ISM is implemented.

On balance, repair processes have been improved during ISM-X, as evidenced by decreases in both man-hours and overall variability of the repair resources of time, people, and parts. However, repair of a number of lines reflects worse performance during ISM-X than during the baseline period. These cases need to be examined individually for relevant characteristics and the problems (e.g., unavailability of parts, capacity, skills) resolved. The substantial number of lines whose repair processes improved can provide a source of

---

[46]Our conversations with ISM-X personnel indicate that this suggestion is being made from all quarters.

[47]Lionel A. Galway and Christopher H. Hanks, *Data Quality Problems in Army Logistics: Classification, Examples, and Solutions,* Santa Monica, CA: RAND, MR-721-A, 1996.

[48]Reportedly, the scrutiny of data from the LSMMs by the RSMM staff has greatly improved both the quality of the data and the consciousness of the importance of maintenance data to maintenance management.

potential solutions. Days in COE should be added as a performance parameter in ISM management and should be a component of a COE's bid.

The shortfall in unserviceable returns is an important problem for the Army as a whole, and it is critical to the resource planning that is one of the primary goals of the ISM concept. The causes during ISM-X need to be further evaluated and solutions found (for example, if units are holding unserviceables, the various ISM managers need to have good visibility of the problem).

Repair reliability and readiness are key issues for any new maintenance concept and need to be monitored closely as ISM is implemented. We encourage the use of serial-number tracking as one way of measuring repair reliability.

The bid process for COE work is a key component of the ISM concept. However, analysis both of the bids relative to historical experience and performance relative to the bid showed surprising variation. To some extent, the former variation may be due to the use of legacy data for the historical baseline (i.e., the man-hours and parts costs may not be accurate or the data may include jobs that were not repairs). However, ISM managers at all levels will need to be able to realistically assess new bids in relation to historical performance. The RSMM is addressing performance-to-bid variation by identifying whether a COE has met its commitment; part of the performance variation relative to bids may be due to the learning curve as COEs develop experience with new repair processes and new methods of doing business.

Finally, conversations with people across the ISM community indicate broad agreement on the need to improve data quality. Actions include improving connectivity among information systems to reduce manual input, developing automated checking of key data elements, and continuing and enhancing feedback to legacy-system users on their input. Most important, the ISM management system needs to have permanent correction of all maintenance data for long-term analysis of Army ISM maintenance.

4. COSTS AND ECONOMIC BENEFITS OF ISM

The potential economic benefits of ISM have commanded much attention during both the ISM PoP and the ISM-X demonstration—perhaps a disproportionate amount, considering that the initial intention of ISM was to improve maintenance support during both peacetime and contingencies. It is important to forecast these benefits as accurately as possible (given the inherent variability of demand for repairs), because they are being used to make prospective reductions in Army logistics budgets. If the forecasts exaggerate the benefits, the Army could lose the money without gaining the compensating reductions and might not be able to achieve the same level of readiness on a reduced budget.

Thus, any estimate of cost avoidance needs to make the best possible use of the information available from the ISM PoP and ISM-X. Some draft cost-benefit analyses estimated Armywide cost avoidance by extrapolating from the ISM PoP data gathered from the participating installations. However, the methodology employed used a constant-percentage cost avoidance across all items. As the data in this section show, considerable variation occurs in productivity and cost avoidance across items, and thus an item-by-item extrapolation should yield improved estimates. Furthermore, the methodology equates cost avoidance at the installation level with cost avoidance for the Army as a whole.

However, cost avoidance to the Army as a whole on a given item differs from that achieved at an installation. A commander at an installation retains OMA dollars if the cost of an ISM repair is less than the retail price. But the Army benefits only if the cost to repair or replace the item at the wholesale level is less than the cost to repair it at the installation. If the local cost avoidance exceeds the Army's, which can happen because AMDF prices and retail credit rates include a management surcharge and because retail credit rates average repair costs across entire Materiel Categories, the Army stock fund suffers a net loss. Thus, estimates of economic benefits need to focus at the Army level.

Finally, the estimates use labor rates that assume costs change linearly with maintenance activity. But certain fixed costs do not vary linearly with activity, and these should be removed when estimating cost avoidance.[49]

---

[49]Although fixed costs may be attributed to labor rates to determine the rates that would ensure full cost recovery, this attribution must be based on an estimation of the total number of labor hours spent on maintenance. Therefore, when estimating the cost avoidance due to changes in workload, which involve changes in the number of labor hours spent on maintenance, fixed costs should be omitted from hourly labor rates.

In addition to these methodological issues, some potential additional costs need to be considered when projecting economic benefits. These include any expansion costs associated with ISM work, any additional personnel required by ISM, and any development and fielding costs associated with the EMIS. This section examines all these issues, beginning by describing where ISM benefits should occur in principle, and then examining the practice of ISM. The section concludes with some suggestions for improving the ISM cost-benefit analysis.

ECONOMIC BENEFITS OF ISM: THE PRINCIPLE

If ISM operates as intended, it should provide economic benefits to the Army in two ways.

First, it should make the GS repair processes more efficient. The processes at a given installation should be more efficient, and the system should direct the repairs to the most efficient place to repair them. Part of the increased efficiency should result from specialization of both skills and parts inventories. Greater specialization of the labor force should enable workers to make repairs more efficiently, so fewer workers are needed to complete the existing workload, or more repairs can be completed by the existing workforce. Furthermore, the stock of parts can be focused on the items repaired at each installation. Another part of the efficiency results from repairing to an Inspect-and-Repair-Only-As-Necessary standard. This more-focused IRON standard should reduce the consumption of spare parts. Moreover, this standard, coupled with the improved efficiency that results from competition and higher repair rates, should also lower the labor hours per repair.[50]

Second, ISM should enable the Army to reduce inventories of components and repair parts at the installation and wholesale levels. As repairs concentrate at COEs, increased specialization and a more consistent flow of unserviceable components should allow installations to reduce their parts inventories. An installation can eliminate most or all of its parts for reparables sent to other COEs, selling them to the installations that use them. Faster turnaround time should also allow installations to reduce their inventories of reparable components, particularly if a repair-and-return-to-stock process is implemented. Finally, improved visibility of reparable components at the wholesale level should permit further inventory reductions. However, to realize this benefit, managers would need access to an improved and expanded EMIS, including information about stock levels at the COEs.

---

[50] As we discuss in greater detail in Section 6 of this report, additional data on component performance would need to be collected to determine whether the IRON standard is more cost-effective on a line-by-line basis. If the IRON standard results in more-frequent repairs for some lines, both repair and transportation costs could increase.

To gain the potential economic benefits of ISM, the Army must make structural changes to reduce costs in one or more of four areas, some of which will require difficult decisions. These areas include reducing the workforce (government or contracted), closing maintenance facilities, reducing procurement, and lowering readiness.

First, if ISM increases overall repair efficiency and worker productivity, for a fixed number of repairs it should be possible to reduce the number of personnel involved in maintenance activities.[51] When direct labor is reduced at an installation, it may also be possible to reduce certain components of the indirect and overhead costs. Some trade-offs may be necessary to retain a capability to meet surge demands generated by a contingency.

Second, if the workload at any particular maintenance organization falls sufficiently as a result of repair-activity consolidation, it may be possible to close some of that organization's facilities and redistribute repair and test equipment. Because increased repairs at the installation level may reduce GS overflow to depots, the least-efficient facilities at either the installation or wholesale level might be closed.

Third, savings may be realized by reducing procurement of new items, either components or repair parts. Such reductions would result from the lower inventory levels created by improved repair processes, better inventory management, and the extended life of repaired components. However, this source of savings has potential implications for the frequency and cost of repairs if the extended life of components causes them to fail more often as they age.

Finally, decreasing readiness directly, rather than decreasing costs for a given level of readiness, is a fourth way to lower maintenance costs. If improved performance cannot compensate for reduced budgets, the Army could simply decide to maintain its equipment at a lower level of readiness, which would reduce the demands on the system. The Army already establishes different levels of readiness for different units. However, we note that degradations in readiness cannot be taken lightly, because the Army's primary job is national defense.

ESTIMATING THE ECONOMIC BENEFITS OF ISM: THE PRACTICE

In practice, many of the Armywide economic benefits of ISM can be achieved only in the long term. Therefore, short-term tests and demonstrations conducted at a few sites, such as the ISM PoP and ISM-X, cannot be used to measure future Armywide economic benefits directly. Data collected during the demonstrations, however, can be used as a basis for

---

[51]Alternatively, a constant number of repair personnel can accomplish an increased number of repairs, which could result in reduced procurement rather than reduced personnel.

estimating potential Armywide economic benefits from ISM.[52] Because the Army is using these estimates to reduce future OMA budgets, it is particularly important that the estimates be as realistic as possible. In this section, we discuss some of the issues involved in estimating ISM costs and benefits and how these estimates might be improved.

The economic benefits measured during the ISM PoP and ISM-X differ somewhat from those that might be achieved in principle. During the ISM PoP and ISM-X, OMA-funded organizations had two ways to avoid costs. First, they avoided OMA costs by using fewer man-hours and fewer parts to repair items that were previously repaired at the installation. Second, they avoided OMA costs by repairing items that previously had been purchased from the wholesale logistics system.[53] The economic benefits of Armywide implementation of ISM have been estimated elsewhere by extrapolating this OMA cost avoidance to all CONUS installations.[54]

The cost avoidance for OMA-funded organizations does not necessarily translate into cost avoidance for the Army as a whole, however, because the Army must also consider ISM's financial effects on stock-funded wholesale logistics organizations performing supply management and depot maintenance. We believe that the Armywide perspective is the appropriate one for estimating the economic benefits of ISM. The recommendations discussed below are directed toward developing improved estimates of Armywide economic benefits that are possible using ISM PoP and ISM-X data.

Extrapolating from ISM PoP and ISM-X Data

One of the methodological issues in estimating the economic benefits of ISM is how to extrapolate from the ISM PoP and ISM-X data to the rest of CONUS. The draft cost-benefit

---

[52]As part of RAND's evaluation of ISM-X, we examined two draft cost-benefit analyses and Computer Systems Development Corporation, *Integrated Sustainment Maintenance Proof of Principle Final Evaluation Report*, October 1994. Since the draft cost-benefit analyses have not been approved by the Army, we do not cite them explicitly, but we discuss their methodology and suggest areas for improvement.

[53]ISM reduces GS overflow by allowing installations to repair more field-level reparables (FLRs) rather than purchasing them from the wholesale system, or to perform more GS-level tasks on depot-level reparables (DLRs) rather than returning carcasses to the wholesale system in exchange for a serviceable replacement. A number of the ISM repair items are classified as DLRs. This does not necessarily mean that the installations are performing depot-repair tasks. The categorization of repair location is far from precise, because some installations have authority to perform depot-level repair tasks, some depots repair FLRs because of a lack of installation capability or capacity, and some repair tasks on DLRs are authorized to be performed at the installation. Indeed, for the ISM items, the vast majority of the repairs on DLR items were authorized as field-level repairs.

[54]For example, one of the draft cost-benefit analyses estimated the economic benefits of ISM in constant dollars over 10 years. Approximately one-third of those costs were attributed to lower maintenance costs, and two-thirds to reduced purchases from the wholesale logistics system.

analyses used constant-percentage cost-estimating factors for repair and requisition costs across all lines. However, productivity gains and cost avoidance varied greatly across lines. In addition, the actual number of items repaired during the ISM PoP and ISM-X did not always meet planned repairs (see the discussion in Section 3). Therefore, we recommend that the economic benefits should be extrapolated on a line-by-line basis rather than as constant factors across all lines, depending on repairs completed as a percentage of demands and cost avoidance on each item. Extending the baseline period or collecting additional repair data as ISM evolves can also help refine estimates of future economic benefits. We discuss each of these issues in greater detail below.

Variation in productivity gains. Changes in productivity varied greatly among the ISM PoP lines, as well as among those added during ISM-X. First, we consider the 65 lines included in the ISM PoP. Table 4.1 shows the direct labor hours required for the ISM PoP repairs relative to the baseline labor hours that would have been required to make the same number of repairs, based on mean labor hours for the period before the ISM PoP. This information has been generated from the data used in the evaluation of the ISM PoP.[55] The 65 ISM PoP lines divide into two groups, based on whether productivity improved during the PoP. Table 4.2 shows the same information, but divides the 65 PoP lines between the 10 with the greatest improvement in productivity and the remaining 55.

Table 4.1

Productivity Changes During ISM PoP

Change in Man-Hours	Number of Lines	Baseline Man-Hours	ISM PoP Man-Hours	Percentage Reduction in Man-Hours
Reduction	34	20,716	14,868	28.2
Increase	24	3,714	5,130	−38.1
No data	7	0	0	0.0
Total	65	24,429	19,997	18.1

NOTE: Percentage reductions are calculated as (ISM PoP Man-Hours – baseline Man-Hours) / baseline Man-Hours. The percentages do not sum vertically because they are weighted by the number of Man-Hours spent on each group of lines. Percentage reductions shown in subsequent tables are calculated in a similar manner.

---

[55]See Computer Systems Development Corporation (1994). We report the ISM PoP data separately, because they have been used to generate the draft cost-benefit analyses. Cost-reduction factors derived from the ISM PoP data were used to extrapolate to ISM-X items in one of the cost-benefit analyses.

Table 4.2

Productivity Changes for Best Lines in ISM PoP

Group	Number of Lines	Baseline Man-Hours	ISM PoP Man-Hours	Percentage Reduction in Man-Hours
Best	10	12,187	7,633	37.4
Remainder	55	12,242	12,365	−1.0
Total	65	24,429	19,997	18.1

As the tables show, the significant reduction in man-hours (or corresponding improvement in productivity) does not come from uniform improvement across all lines, but is concentrated in a small number of lines. Although 34 lines show reduced man-hours, 24 show an increase, and 7 had no repairs at all. Moreover, the 10 key lines from which most of the improvement comes show a productivity improvement of 37.4 percent, whereas the remaining 55 lines as a group show a net reduction in productivity of 1.0 percent (Table 4.2).

Tables 4.3 and 4.4 show the corresponding results for the 91 lines that were added to ISM-X at the fifth and sixth PP&C meetings. These tables show more-positive results, but they have much the same pattern as the PoP data. The measure of overall labor productivity improved to 27.6 percent, but again the improvement is distributed unevenly. As Table 4.4 demonstrates, only 10 of the 91 additional lines in the demonstration accounted for the majority of the labor hours and the corresponding improvement in repair productivity.

These results indicate that productivity gains during the ISM PoP and ISM-X were highly variable and tended to be driven by a relatively small set of lines. Furthermore, the ISM PoP lines were chosen not to be a representative sample of GS reparables but either to be candidates with the highest potential cost avoidance or to balance workloads across installations. Accordingly, estimates of the economic benefits of ISM could be improved by making extrapolations from the data on a line-by-line basis or, at least, grouped by item type, such as engines, transmissions, etc.

Variation in cost avoidance. Tables 4.5 and 4.6 show that repair cost avoidance (as measured by the draft cost-benefit analyses) also varied greatly among lines during the ISM PoP.[56] The tables compare the cost of the ISM PoP repairs with the baseline cost of performing the same number of repairs before the ISM PoP.[57] From Table 4.5 it is clear

---

[56]These tables are based on data in Computer Systems Development Corporation (1994).

[57]For comparability with the draft cost-benefit analyses, pre-ISM costs are based on fully burdened labor rates and repair parts costs. The ISM PoP and ISM-X repair costs also include the additional costs of packing (if applicable) and transportation. We discuss the problems involved with using fully burdened labor rates below.

Table 4.3

Productivity Changes During ISM-X

Change in Man-Hours	Number of Lines	Baseline Man-Hours	ISM-X Man-Hours	Percentage Reduction in Man-Hours
Reduction	42	26,723	18,581	30.5
Increase	19	1,427	1,797	−26.0
No data	30	0	0	0.0
Total	91	28,150	20,393	27.6

Table 4.4

Productivity Changes for Best Lines in ISM-X

Group	Number of Lines	Baseline Man-Hours	ISM-X Man-Hours	Percentage Reduction in Man-Hours
Best	10	21,739	14,493	33.3
Remainder	81	6,411	5,900	8.0
Total	91	28,150	20,393	27.6

that almost equal numbers of lines showed increased repair costs (28) as reduced repair costs (30) during the ISM PoP. Table 4.6 shows that four lines drove the majority of repair cost avoidance, and that the four lines together had cost avoidance that exceeded the total for all 65 lines. For the majority of lines that were neither the four best nor four worst, the reduction in overall repair cost (including transportation) was about 6 percent of the baseline repair costs. Note that this comparison implicitly includes any potential change in repair standards or the difficulty of repairs that could occur under ISM.

Tables 4.7 and 4.8 show similar variation among the 91 prime lines that were added to ISM-X at PP&Cs 5 and 6. For these components, a much higher fraction of the new lines showed repair-cost reductions. Note, however, that almost a third of the lines had no repairs at all during ISM-X, and that only 6 lines generated most of the cost avoidance. For the remaining 85 (combining the second and third rows of Table 4.8), the net reduction in repair costs was about $230,000, or roughly 12.5 percent of the baseline cost. For 22 of the lines, the calculated repair costs were higher under ISM-X than before the demonstration, but the majority of the higher costs were derived from only 8 of these lines.

Table 4.5

ISM PoP Repair Cost Avoidance

Change in Repair Cost	Number of Lines	Baseline Cost	ISM PoP Cost	Cost Avoidance	Percentage Cost Avoidance
Reduction	30	$1,609,524	$1,213,208	$396,317	24.6
Increase	28	$442,483	$606,787	($164,304)	−37.1
No repairs	7	$0	$0	$0	0
Total	65	$2,052,007	$1,819,995	$232,012	11.3

Table 4.6

ISM PoP Repair Cost Avoidance for Best and Worst Lines

Group	Number of Lines	Baseline Cost	ISM PoP Cost	Cost Avoidance	Percentage Cost Avoidance
Best	4	$795,866	$534,870	$260,996	32.8
Worst	4	$239,108	$331,869	($92,761)	−38.8
Remainder	57	$1,017,033	$953,256	$63,777	6.3
Total	65	$2,052,007	$1,819,995	$232,012	11.3

These tables indicate that ISM-X did generate productivity increases and significant total repair-cost avoidance. They also show that some lines seem to be more suitable for regional repair than others. Given the variation in cost avoidance among lines, using a constant-percentage cost-estimating factor across lines may yield a misleading estimate of cost avoidance. An improved estimate of the economic benefits of ISM could be obtained by extrapolating cost avoidance on a line-by-line basis.

Actual versus planned repairs. As discussed in the preceding section, the actual number of items returned to COEs often fell below the planned number during the ISM-X demonstration. Therefore, it is important to use the actual number of repairs achieved by the COEs as a percentage of demands, rather than the planned 80 percent of demands, to extrapolate cost avoidance to the rest of CONUS. Although the planned number of repairs may be achievable in the long term, it is also possible that, because of constraints on labor and facilities, the planned number may not be attained. The ratio of actual to planned repairs varied across lines, so that ratio also should be extrapolated on a line-by-line basis.

Table 4.7

ISM-X Repair Cost Avoidance

Change in Repair Cost	Number of Lines	Baseline Cost	ISM-X Cost	Cost Avoidance	Percentage Cost Avoidance
Reduction	39	$3,193,972	$1,957,494	$1,236,478	38.7
Increase	22	$441,000	$498,293	($57,293)	−13.0
No repairs	30	$0	$0	$0	0
Total	91	$3,634,972	$2,455,787	$1,179,185	32.4

Table 4.8

ISM-X Repair Cost Avoidance for Best and Worst Lines

Group	Number of Lines	Baseline Cost	ISM-X Cost	Cost Avoidance	Percentage Cost Avoidance
Best	6	$1,795,201	$848,962	$946,239	52.7
Worst	8	$227,997	$276,107	($48,110)	−21.1
Remainder	77	$1,611,774	$1,330,718	$281,056	17.4
Total	91	$3,634,972	$2,455,787	$1,179,185	32.4

Extending ISM and baseline data. The estimated economic benefits for ISM PoP lines are calculated relative to the nine months of baseline data collected for the period before the ISM PoP. These data are acknowledged to contain numerous errors and omissions, and are frequently derived from a relatively small number of repairs. Table 4.9 shows that the mean number of repairs per line in the PoP baseline data exceeds 25, but only 19 of the 65 lines actually had that many repairs. While these 19 lines accounted for over 75 percent of the total repairs, 28 had 10 or fewer repairs from which to derive the baseline repair-cost information.

Table 4.10 shows the same information for the lines added at PP&Cs 5 and 6 during ISM-X. In this case, the baseline data represent repairs made between October 1, 1994, and June 30, 1995. Again, a significant number of lines (25 of 90) have five or fewer repairs in the baseline, and 42 lines (over 45 percent) have 10 or fewer repairs. The vast majority of the total repairs (83 percent) come from only 26 lines averaging over 100 repairs each.

- 42 -

Table 4.9

Number of Repairs for Pre-ISM PoP Baseline

Repair Group	Lines	Total Repairs	Mean Repairs
0	1	0	0
1–5	14	38	2.7
6–10	13	95	7.3
11–25	18	284	15.8
>25	19	1,252	65.9
Total	65	1,669	25.7

Table 4.10

Number of Repairs for Pre-ISM-X Baseline

Repair Group	Lines	Total Repairs	Mean Repairs
0	2	0	0
1–5	23	56	2.4
6–10	17	136	8.0
11–25	22	389	17.7
>25	26	2,873	110.5
Total	90	3,454	38.4

The tables illustrate that the baseline repair data for the ISM PoP and ISM-X may not provide a strong basis for extrapolation of cost avoidance. Furthermore, the ISM PoP and ISM-X covered only limited time periods and were hampered by slow startups and other problems (including the federal budget problems during ISM-X), which limited the number of repairs for many lines. Measurements of the future long-term economic benefits of ISM could potentially be improved by extending baseline data over a longer period and by collecting additional data on repairs and repair costs as ISM evolves.

Prices and Cost Factors

Another important methodological issue in estimating the economic benefits of ISM is the choice of prices and cost factors used to evaluate changes in repair and requisition costs. If the goal is to estimate the changes in costs to the Army as a whole as a result of Armywide implementation of ISM, then prices and cost factors should be chosen to reflect such changes as realistically as possible. The draft cost-benefit analyses use retail stock-fund prices and credit rates, and fully burdened labor rates, to estimate the economic benefits of ISM. As we discuss below, using these rates to evaluate repair and requisition costs may overestimate the potential Armywide cost avoidance available from ISM, particularly in the short run. We recommend that price and cost factors be adjusted to reflect more realistically the avoided costs to the Army as a whole.

Use of retail prices and credit rates. From the perspective of OMA-funded units, if installation-level GS maintenance facilities can repair items previously bought from the wholesale system, the installation saves the difference between the local repair cost and the retail price of a replacement. The retail price is the AMDF price[58] for field-level reparables, or the AMDF price minus the retail credit for depot-level reparables, if the unit returns an unserviceable carcass.

We define retail credit rates as the credits that OMA customers receive from the retail stock fund when unserviceable DLRs are returned to the wholesale system.[59] They are set as the same percentage of AMDF price for all items in a Materiel Category. Actual retail credit rates for the fourth quarter of FY95 by Materiel Category are shown in Table 4.11. They are set quarterly to balance with the wholesale credits received by the retail stock fund, which vary on a line-by-line basis, based on the cost of repairing or replacing the specific item.

Table 4.11

Retail Credit Rates, 4th Quarter FY95

Materiel Category	MSC	Credit as Percentage of AMDF Price
B	ATCOM Troop Support	48.2
G	CECOM Electronics	45.2
H	ATCOM Aviation	50.5
K	TACOM	48.1
L	MICOM	54.8
M	ACALA	47.6
U	CECOM Communications	45.2

SOURCE: Published quarterly by HQDA.

---

[58]The AMDF price is the acquisition cost of the item plus the wholesale Supply Management surcharge, which is usually in the range of 20–30 percent.

[59]If an OMA customer returns an unserviceable DLR, and there is no repair program on the installation for that item, the customer receives credit based on the "alternate credit table" for DLRs, as shown in Table 4.11. The retail credit rates received for DLRs when they are returned to the wholesale system should not be confused with the credit rates for Reparable Exchange (RX) items, which are repaired at the installation level. OMA-funded units receive RX credit rates of 80–90 percent of the AMDF price when they return unserviceables that are repaired in ISM.

To summarize, the retail price is 100 percent of the AMDF price for FLRs, and 45–55 percent of the AMDF price for DLRs, if the item cannot be repaired locally and the unserviceable carcass is returned to the wholesale system. If the costs of ISM repairs are lower than the retail price, then the unit commander retains OMA dollars, which can be spent for other purposes. However, from the Army's perspective, cost avoidance is represented by the difference between the cost of repairing or replacing the item at the wholesale level and the cost of performing the repair in ISM.[60] If local OMA cost avoidance exceeds the Army's, it represents a drain on the Army's stock funds and will have to be recovered in future-year budgets.[61]

Retail prices differ from the Army's cost avoidance for two reasons. First, AMDF prices and retail credit rates include the Supply Management surcharge, which is intended to recover the overhead costs of the wholesale supply system. If ISM performs additional repairs, the wholesale system experiences fewer transactions, so it recovers less of its overhead costs. The wholesale system may reduce some direct transaction costs, but it cannot reduce its fixed costs unless it reduces personnel or facilities. Thus, the Army as a whole does not avoid the full costs represented by the Supply Management surcharge when transactions are reduced.

Second, retail credit rates average the cost of repairs (adjusted for washouts) across entire Materiel Categories. There are only one or two Materiel Categories for each of AMC's Major Subordinate Commands (MSCs), so, for example, all TACOM items have the same retail credit rate as a percentage of AMDF price. In practice, however, wholesale repair costs are not a constant percentage of AMDF price. Some lines have higher-than-average repair costs, and others have lower. The Army's cost avoidance is assessed by comparing the wholesale repair or replacement costs with ISM repair costs on a line-by-line basis. Furthermore, since ISM lines are chosen on the basis of the retail price paid by OMA-funded units rather than avoided costs at the wholesale level, ISM may not be targeting the most cost-effective repairs for the Army as a whole.

To illustrate this problem, Table 4.12 compares local estimated OMA cost avoidance per item based on the retail price paid by OMA customers with the Army's estimated cost

---

[60]The Army's cost avoidance depends on what would have happened if an item had not been repaired in ISM but instead had been purchased from the wholesale system, and a carcass returned, if applicable. For example, if the carcass would have been overhauled, the cost of the overhaul is the avoided cost.

[61]This difference occurs because wholesale surcharges are set to recover overhead costs as a percentage of the total expected value of transactions (for Supply Management) or the total expected number of direct labor hours (for Depot Maintenance).

Table 4.12

Local Versus Army Cost Avoidance per Line

FSC	Prime NIIN	Line	Total ISM Cost[a]	Retail Price	Wholesale Cost	Local Cost Avoidance	Army Cost Avoidance
ATCOM Lines							
2840	01-013-1339	T-63 Aircraft Engine	$43,584	$71,144	$65,107	$27,560	$21,523
2840	01-030-4890	CH-47 Aircraft Engine	37,844	310,486	149,768	272,642	111,924
2840	01-070-1003	UH-60 Aircraft Engine	139,158	287,270	146,426	148,112	7,268
1615	01-106-1903	UH-60 Blade	2,295	41,185	15,870	38,889	13,575
1615	01-113-8188	UH-60 Blade Rotary	2,256	22,175	15,126	19,919	12,870
2840	01-114-2211	AH-64 Aircraft Engine	121,072	266,868	144,558	145,796	23,486
1615	01-137-8137	OH58A/C Blade	438	6,683	6,075	6,245	5,637
1240	01-232-6568	Avionics Day Sensor	16,214	71,218	12,834	55,004	−3,380
1615	01-254-7793	AH-64 Gear Box	6,572	18,253	15,295	11,681	8,723
1615	01-270-2982	UH-1 Hub Rotor	4,960	25,273	19,015	20,313	14,055
1560	01-301-8212	UH-60 Horizontal Stabilizer	3,134	26,236	15,753	23,102	12,619
1270	01-307-9447	Avionics Turret Sensor Sight	595	94,722	11,254	94,127	10,659
1615	01-312-2387	AH-64 Blade Tail	1,609	7,062	1,376	5,453	−233
1615	01-332-0702	AH-64 Blade Rotary Wing	15,920	43,847	51,217	27,927	35,297
CECOM Lines							
5895	00-457-0571	Tactical Communications Receiver/Transmitter	237	339	154	102	−83
5955	00-853-5915	Tactical Comm. Oscillator	113	108	69	−5	−44
7025	01-237-8098	Avionics Display Unit	8,356	11,213	5,137	2,857	−3,219
5841	01-245-9091	Avionics Altimeter Receiver/Transmitter	3,838	6,298	5,206	2,460	1,368
5820	01-266-5964	Radio Set	178	484	556	306	378
MICOM Lines							
1240	01-096-5151	TOW M2/3 Sight Assembly	65,877	55,237	5,254	−10,640	−60,623
1270	01-259-0154	AAH MICOM Electronic Unit	1,998	5,227	2,689	3,229	691
1055	01-338-1703	MLRS Payload Interface	17,970	27,171	13,980	9,201	−3,990
TACOM Lines							
2590	00-753-8687	M35A2 Winch Drum[b]	555	1,974	1,577	1,419	1,022
2530	01-091-1659	M939 Axle Assembly	3,646	5,991	2,105	2,345	−1,541
2520	01-161-2136	HMMWV Transmission	788	2,041	1,300	1,253	512
2510	01-179-7523	M1 Operator's Panel	1,312	2,334	1,146	1,022	−166
2815	01-260-0212	M109A5 Engine	4,836	11,023	14,564	6,187	9,728
4140	01-284-5722	M109 Vaneaxial Fan	664	610	938	−54	274
2815	01-295-7458	M113A2 Diesel Engine[b]	3,170	10,039	8,018	6,869	4,848
2815	01-314-7940	HMMWV Engine	2,550	6,266	3,094	3,716	544

[a]*Total ISM cost* is defined as the winning bid plus average washout, transportation, and packing costs per item. It was obtained from spreadsheets distributed at PP&C 6 by dividing the minimum total cost associated with the winning bid by the expected number of demands.

[b]These lines receive zero unserviceable credit from the MSCs, indicating that they are FLRs and/or that repair is uneconomical at the wholesale level.

avoidance per item based on wholesale repair or replacement costs. Both local and Army cost avoidances are estimated relative to the total costs per item associated with the winning bids that were selected at PP&C 6. The total costs include parts, fully burdened labor costs, washout costs, and packing and transportation costs. We focus on PP&C 6 because at earlier PP&Cs, information on washouts, transportation, and packing costs was not systematically combined with bids, which cover only parts and labor.[62] Retail prices are AMDF prices for FLRs and AMDF prices minus retail credits for DLRs.[63]

We derive wholesale costs from wholesale serviceable and unserviceable credit rates obtained from the respective MSCs.[64] Taking the difference between serviceable and unserviceable credit rates eliminates the Supply Management surcharge and the averaging of repair costs across Materiel Categories, and leaves a wholesale overhaul cost adjusted to account for the cost of replacing washouts. When items are not repaired at the wholesale level (when items are FLRs or repairs are uneconomical, for example), the wholesale cost is the replacement cost.

Because the Supply Management surcharge is removed, the wholesale costs are likely to be lower on average than the retail prices. The averaging effect can work in either direction, so results can differ when comparing estimated local cost avoidance with the Army's. In most cases, projected ISM costs are lower than wholesale costs, but the Army's cost avoidance is not as high as the local one. For a few lines, wholesale costs are higher than the retail price, so the Army's cost avoidance is higher than local cost avoidance. However, there are also some items for which projected ISM costs were higher than wholesale costs, indicating a potential for higher total costs under ISM than if the lines had been repaired (or replaced) at the wholesale level. In a few cases, COE lines were awarded

---

[62]According to materials distributed at PP&C 6, the total cost is defined as

$$[\text{total bid\{regional demands\}}(1 - \text{washout rate})] +$$
$$[(\text{regional demands})(\text{washout rate})(\text{washout cost})] +$$
$$[(\text{regional demands})(\text{washout rate})(\text{buy price})] +$$
$$[(\text{regional demands} - \text{local demands})\{2(\text{transport cost} + \text{packing cost})\}].$$

The total bid is labor hours multiplied by the fully burdened labor rate, plus parts costs.

[63]The items shown in Tables 4.12 and 4.13 represent all the COE lines awarded at PP&C 6 that are managed by ATCOM, CECOM, MICOM (U.S. Army Missile Command), or TACOM, except the AH-1 engine, for which ATCOM did not provide credit information. ACALA (U.S. Army Armament and Chemical Acquisition and Logistics Activity) was not able to respond to our request for information on wholesale credit rates.

[64]The wholesale serviceable credit rate is equal to the acquisition cost (AMDF price – wholesale surcharge). The wholesale unserviceable credit rate is equal to

$$\text{AMDF price} - \text{wholesale surcharge} - \text{repair cost},$$

adjusted for the replacement cost of washouts if the item is a DLR; it is zero if the item is an FLR.

even though total projected ISM costs were higher than the retail price. This situation can occur if the item is being repaired by ISM to improve readiness.[65]

Examples from Table 4.12 illustrate the two cases. The first line, the T-63 aircraft engine, demonstrates the typical case: an ISM cost lower than wholesale, but Army savings not as high as local savings. The ISM cost of $43,584 is lower than the retail price of $71,144, for a local cost avoidance of $27,560. However, the Army's cost avoidance is only $21,523, or over $6,000 less. The AH-64 rotary-wing blade demonstrates the second case: Army cost avoidance exceeds that of the installation. Here, the Army avoidance is over $7,000 higher than that of the installation ($35,297 versus $27,927). But for the avionics day sensor repair, the installation avoids $55,004 in costs while the Army actually loses $3,380.

Table 4.13 shows how these differences in the projected cost avoidance per item can influence overall ISM cost-avoidance estimates. The local or Army cost avoidance per item is multiplied by an estimated number of demands diverted from the wholesale system for each item. Our estimate is based on an increase of repairs under ISM from 40 percent of annual Central Demand Data Base (CDDB) demands to 80 percent of CDDB demands.[66] For this set of items (which is not necessarily a representative sample), local cost avoidance based on retail prices totals $8.5 million, but the Army's cost avoidance based on wholesale repair or replacement costs is $3.0 million.

As this example illustrates, when items are repaired in ISM rather than returned to the wholesale system, retail prices do not represent avoided costs to the Army as a whole. To improve estimates of the economic benefits of ISM, cost factors should be chosen to reflect costs that change at the wholesale level as more items are repaired in ISM.

Although the wholesale repair or replacement costs used in Tables 4.12 and 4.13 are better than retail prices as estimates of avoided costs at the wholesale level, they are not ideal. Some reflect depot-repair programs that are not currently active and that could not be restarted at the same cost. Furthermore, both ISM costs and depot repair costs are based on fully burdened labor rates, which may include overhead costs that do not vary with the

---

[65]It can also occur if parts and labor costs are less than 80 percent of the retail price, but washout and transportation costs cause the total cost per item to exceed the retail price. Under the rules for granting COE status, the total cost to repair (including transportation and packing) must be less than 80 percent of the net cost to purchase. See Section 6 for the other selection rules.

[66]Since installations were repairing approximately 40 percent of annual CDDB demands for COE items before ISM, and projected ISM total costs are based on repairing 80 percent of annual CDDB demands (for the year prior to the PP&C), this fraction represents an approximation of the potential number of annual CDDB demands diverted from the wholesale system by ISM. The costs of washouts (including NRTSs for DLRs) are deducted from cost avoidance on diverted demands.

number of repairs and, therefore, are not good estimates of actual changes in costs. We discuss this issue in greater detail in the next subsection.

Table 4.13
Total Local Cost Avoidance Versus Total Army Cost Avoidance

FSC	Prime NIIN	Line	Diverted Demands	Local Cost Avoidance	Army Cost Avoidance
ATCOM Lines					
2840	01-013-1339	T-63 Aircraft Engine	12	$330,725	$258,277
2840	01-030-4890	CH-47 Aircraft Engine	1	272,642	111,924
2840	01-070-1003	UH-60 Aircraft Engine	4	592,448	29,071
1615	01-106-1903	UH-60 Blade	13	505,563	176,474
1615	01-113-8188	UH-60 Blade Rotary	8	159,355	102,963
2840	01-114-2211	AH-64 Aircraft Engine	3	364,490	58,714
1615	01-137-8137	OH58A/C Blade	16	99,919	90,199
1240	01-232-6568	Avionics Day Sensor	8	440,029	−27,040
1615	01-254-7793	AH-64 Gear Box	5	58,405	43,614
1615	01-270-2982	UH-1 Hub Rotor	10	203,129	140,551
1560	01-301-8212	UH-60 Horizontal Stabilizer	2	46,205	25,238
1270	01-307-9447	Avionics Turret Sensor Sight	11	988,339	111,923
1615	01-312-2387	AH-64 Blade Tail	31	166,319	−7,109
1615	01-332-0702	AH-64 Blade Rotary Wing	16	446,820	564,746
CECOM Lines					
5895	00-457-0571	Tactical Communications Receiver/Transmitter	222	22,597	−18,508
5955	00-853-5915	Tactical Comm. Oscillator	252	−1,314	−11,112
7025	01-237-8098	Avionics Display Unit	7	18,568	−20,920
5841	01-245-9091	Avionics Altimeter Receiver Transmitter	34	82,416	45,821
5820	01-266-5964	Radio Set	93	28,329	35,000
MICOM Lines					
1240	01-096-5151	TOW M2/3 Sight Assembly	14	−148,956	−848,715
1270	01-259-0154	AAH MICOM Electronic Unit	9	27,450	5,874
1055	01-338-1703	MLRS Payload Interface	3	27,603	−11,971
TACOM Lines					
2590	00-753-8687	M35A2 Winch Drum	43	61,031	43,946
2530	01-091-1659	M939 Axle Assembly	4	9,378	−6,164
2520	01-161-2136	HMMWV Transmission	110	137,781	56,271
2510	01-179-7523	M1 Operator's Panel	82	83,320	−13,559
2815	01-260-0212	M109A5 Engine	35	213,450	335,599
4140	01-284-5722	M109 Vaneaxial Fan	53	−2,792	14,392
2815	01-295-7458	M113A2 Diesel Engine	320	2,194,710	1,549,119
2815	01-314-7940	HMMWV Engine	291	1,079,552	157,940
Total				$8,507,512	$2,992,558

Use of fully burdened labor rates. Fully burdened labor rates for GS maintenance facilities were developed to facilitate cost comparisons across installations and to estimate Armywide cost avoidance. These rates have helped the Army to reach a better understanding of local maintenance costs, and they provide a more direct comparison with fully burdened depot labor rates. However, if the goal is to compare changes in costs at installations with changes in costs at the wholesale level as more lines are repaired in ISM, costs that do not vary with maintenance activity should not be included.

The use of fully burdened rates implicitly assumes that all overhead costs vary linearly with direct labor, so this practice may overestimate actual changes in costs. Thus, unless all indirect and general and administrative (G&A) expenses scale up and down with direct labor, fully burdened labor rates are not good estimates of actual cost avoidance. Even if some overhead costs are expected to fall in the long term as workload is reduced at some facilities, it may take some time for these cost reductions to be realized. Therefore, cost avoidance may be overestimated to a greater degree in the first few years of ISM implementation. In addition, fully burdened rates may include fixed costs that cannot be changed in the short term, or sunk costs[67] that cannot be recovered. One important example is the incorporation of prior-year stock-fund losses into depot hourly rates. Clearly, these losses are sunk costs that cannot be reduced when depot repairs are reduced.

To understand the contribution of indirect and G&A costs to fully burdened labor rates, we examined the cost mapping performed for ISM-X.[68] The breakdown of the fully burdened labor rates into direct, indirect, and G&A costs is shown in Table 4.14. The indirect costs at ISM-X installations range between 50 and 125 percent of the direct labor costs, and average about 90 percent. The breakdown of these costs by installation and category (for the active forces) is shown in Table 4.15. As the table indicates, these costs consist primarily of personnel, general supplies (excluding repair parts), and contracted operations (such as tire recapping, body work, and other maintenance activities).

---

[67] Fixed costs, which do not vary with the level of activity, can be divided into avoidable fixed costs and sunk costs. Avoidable fixed costs would not be incurred if a facility were closed, whereas sunk costs were incurred in the past and cannot be changed by current or future decisions.

[68] Spreadsheets of the cost maps used for ISM-X, showing detailed breakdowns by installation, were provided to RAND by Military Professional Resources, Inc. (MPRI). The development of fully burdened labor rates is described by George C. Ogden, Jr., and David M. Robinson II, in *Development of Full Cost Solutions for the Integrated Sustainment Maintenance Proof of Principle*, Alexandria, VA: MPRI, September 10, 1993.

Table 4.14

FY94 Cost Factors from Cost Mapping ($/hour)

Organization	Direct	Indirect	G&A	Total	G&A as Percentage of Total
Bliss-Aviation	30.03	14.85	5.22	50.10	10.4
Bliss-DOL	22.65	17.90	6.65	47.20	14.1
Carson-DOL	23.51	21.59	7.12	52.22	13.6
Hood-190th GSU	33.59	22.75	3.51	59.85	5.9
Hood-Aviation	19.21	23.81	3.06	46.08	6.6
Hood-DOL	16.27	17.10	3.41	36.78	9.3
Riley-DOL	20.53	22.10	4.74	47.38	10.0
Sill-DOL	19.04	20.49	3.37	42.90	7.9

NOTE: ISM-X cost maps for PP&Cs 5 and 6 were based on FY94 data, which represent the last complete year of installation-level financial data from the Standard Financial System (STANFINS) that was available at the time.

Table 4.15

Indirect Costs from Cost Mapping ($/hour)

Organization	Direct Labor	Indirect Labor	Contracts	Supplies	Other	Total
Bliss-Aviation	30.03	7.20		7.22[a]	0.43	14.85
Bliss-DOL	22.65	8.94	4.17	4.40	0.49	17.90
Carson-DOL	23.51	5.55	5.78	10.26	0.00	21.59
Hood-190th GSU	33.59	19.31	0.25	3.19	0.25	22.75
Hood-Aviation	19.21	7.77	3.62	12.13	0.29	23.81
Hood-DOL	16.27	5.40	3.52	7.67	0.51	17.10
Riley-DOL	20.53	12.92	1.55	7.56	0.07	22.10
Sill-DOL	19.04	8.33	1.90	6.49	3.78	20.49

[a]Contracts and supplies are combined in cost-map data.

Because some of these activities and costs are closely associated with overall maintenance levels, we might expect them to vary proportionately with ISM maintenance activity. However, others may be more closely associated with non-ISM repairs, or they may vary with the number of items repaired, for example, rather than with the number of direct labor hours spent on ISM repairs. Thus, assuming that indirect costs will scale up and down in direct proportion to the number of labor hours spent on ISM repairs may lead to an overestimate of cost avoidance. This may be especially true in the short term, since changes in indirect costs may lag behind changes in direct labor hours.

Table 4.14 indicates that G&A costs range from 6 to 14 percent of the total hourly rates. Table 4.16 shows the total mapped G&A costs for the eight maintenance organizations participating in ISM-X and the fraction of this total contributed by the major categories. Many of these functions, such as command staffing, security, and Morale, Welfare, and Recreation (MWR) activities (included in the "Other" category), do not have any direct relationship to maintenance activities. Others, such as fire prevention or real-property maintenance, may be too "lumpy" to be affected significantly.[69] Relatively small increases or decreases in overall maintenance work forces or workload will not be sufficient to alter total expenditure levels for these functions.

The remaining overhead functions, such as utilities, communications, transportation, and resource management, are likely to vary with the amount of maintenance being performed, but not necessarily with the number of direct-labor hours. Furthermore, all G&A costs are attributed to maintenance activities based on their shares of military, civilian, or total personnel on the installation, or of their shares of the total square feet of building area on the installation. Even the cost of utilities cannot be directly calculated, because most buildings are not separately metered.

Thus, estimates of cost avoidance based on fully burdened labor rates, which assume that indirect and G&A costs scale up and down proportionately with direct labor, may

Table 4.16

Primary Categories of G&A Costs

Category	Allocation Method	Allocated	Percentage of Total
Engineer support	Area	$1,459,339	17.0
Real property maintenance	Area	$1,453,001	17.0
Utilities	Area	$1,397,990	16.3
Personnel operations	Personnel	$1,157,345	13.5
Environmental compliance	Personnel	$438,599	5.1
Supply operations	Personnel	$343,656	4.0
Transportation	Personnel	$300,242	3.5
Fire prevention	Area	$287,067	3.4
Minor construction	Area	$285,974	3.3
Automation activities	Personnel	$276,340	3.2
Other	Personnel	$1,165,086	13.6
Total		$8,564,639	100.0

[69]Lumpy costs can only be incurred in relatively large units. For example, the level of activity on the installation would have to change dramatically for fire prevention to add or cut a fire crew.

include costs that do not vary with maintenance activity, or that vary more directly with factors other than labor, such as parts costs or the number of repairs.[70]

Fully burdened labor rates calculated for depot repairs suffer from the same problems.[71] An improved estimate of cost avoidance could be obtained by omitting indirect and G&A costs that do not vary with maintenance activity, or by attributing them on a more appropriate basis, for both the depots and installations. An improved estimate of the economic benefits of ISM depends on identifying the costs that change with maintenance activity and omitting those that do not.[72]

Inventory Reductions

In principle, the economic benefits of ISM are likely to include reduced inventory in three areas: installation-level stocks of reparable components, because of faster turnaround times; repair-parts inventories at COEs, because of consolidation; and wholesale-level stocks of reparable components, because of decreased demands on wholesale. Although one of the draft cost-benefit analyses attempted to estimate future inventory reductions resulting from ISM, this category was omitted from a later draft cost-benefit analysis.

As ISM evolves, we recommend that inventories of reparable components and repair parts be monitored at the installation and wholesale levels to determine whether the expected benefits are being achieved and to estimate future inventory reductions. Such monitoring should also help to determine whether other policy changes may be needed to achieve greater inventory reductions over the long term.

Possible Additional Costs

Based on our installation visits and attendance at PP&Cs, we believe that the draft cost-benefit analyses may omit or understate some of the costs of implementing ISM. Below, we discuss three areas in which such understatement may occur: expansion of maintenance capacity, ISM personnel requirements, and EMIS development costs.

Expansion of maintenance capacity. The draft cost-benefit analyses are based on the assumption that installation-level repairs can increase from an average of 40 percent of

---

[70]Activity-based costing could help the Army to identify more accurately which costs change with maintenance activity and which costs are fixed.

[71]Depot maintenance rates are set to recover full costs under DBOF policy. But since they include fixed costs, they are not a good basis for estimating cost avoidance due to changes in workload.

[72]The cost maps developed by MPRI are a starting point for identifying those costs that change with maintenance activity. To our knowledge, these cost maps are the first quantitative assessment of how costs vary across installations. The Army needs such information to make cost-effective choices for repair-location decisions.

CDDB demands to 80 percent of CDDB demands for each line, without an increase in maintenance personnel or facilities. This means that productivity on these lines must virtually double, unless repairs are reduced or productivity is also increased on non-ISM lines. If these productivity increases cannot be achieved over the long run, then the higher numbers of ISM repairs can be completed only by incurring additional direct or indirect costs for facilities, personnel, or equipment, or by buying repairs or replacements for non-ISM lines. Therefore, a full cost-benefit analysis of ISM should discuss the potential for additional maintenance-capacity costs or repair costs for non-ISM lines if productivity assumptions are not met.

Evidence from the ISM PoP and ISM-X shown above indicates that, although repair productivity increased, target numbers of repairs were not achieved in the short run. Furthermore, there is some evidence that installations are adding repair capacity. At PP&Cs 6 and 7, the LSMMs discussed both completed and planned investments in facilities and equipment that were being undertaken, at least in part, to make ISM repairs. To the extent that these investments are being driven by ISM, they should be included in ISM costs. Furthermore, if adequate repair capacity already exists at other installations or at the wholesale level, these new facilities and equipment may be redundant. As ISM expands, the RSMMs and NSMM should carefully monitor the overall capacity needed for sustainment maintenance.

ISM personnel requirements. Our observations during the demonstration and discussions with ISM-X participants suggest that ISM (and data entry and analysis using the maintenance management software package EMIS) may create a significant additional workload for installation DOL personnel. The cost-benefit analyses and Department of the Army (DA) policy assume that no incremental personnel authorizations will be needed at the installations, which may not be the case in the long run.

During the ISM PoP and ISM-X, contractor personnel at both the RSMM and LSMMs supported ISM and EMIS activities, including data entry, analysis, and report preparation. The implementation plans for ISM state that each installation will nominally require three people to operate effectively: a logistics management specialist, a supply analyst, and a maintenance analyst. These requirements could be increased or decreased, depending on local troop populations, equipment densities, and other factors. If contractor personnel are not used to staff these positions, as they have been during the PoP and ISM-X demonstration, there is no clear evidence that the installation DOLs can realign or restructure their maintenance operations to perform all this additional work, particularly given the EMIS data entry and ISM paperwork-processing requirements at the COEs.

Moreover, the results of the ISM PoP and demonstration indicate that at some locations it may be difficult, although possible, to "dual-hat" the Installation Maintenance Management Officer (IMMO). Finally, under the current plans for implementing ISM, RSMM personnel will not absorb any LSMM functions.[73]

During the ISM-X evaluation, we also found that ISM can increase workload substantially in the transportation and resource-management operations of the DOL. While implementing ISM as a "repair-and-return-to-stock" system might reduce the resource-management demands (by eliminating the processing of charges against Military Interdepartmental Purchase Requests [MIPRs]), this change will not substantially reduce the transportation workload. Even if items are not returned to the original owner, the COEs must still ship a replacement item from the supply system for each unserviceable return.[74]

In summary, implementation of ISM is likely to improve visibility and efficiency of repair processes, but is also likely to increase the workload within the DOLs. The draft cost-benefit analyses assume that this increased workload can be accommodated solely through realigning and restructuring of existing workforces, without hiring additional contractor or government personnel. The evidence from the ISM PoP and ISM-X indicates that realignment and restructuring may be difficult for many installations to accomplish. As a result, ISM may not operate effectively, or additional funding may be needed for hiring contractor personnel. The approximate annual cost for one additional contracted employee at each of the 33 CONUS installations that will eventually participate in ISM (unless some installations do not become eligible for COEs because of insufficient economies of scale) is between $1 and $2 million.

It may be possible to avoid some of these costs by improving EMIS or its replacement system. For example, the data-entry requirements could be reduced and the data-entry process simplified, either by making EMIS communicate with additional transportation and supply systems or by designing an improved interface for data entry that could incorporate lookup tables, data checking, and other user-friendly features to accelerate the input process and make manual data entry more accurate.

EMIS development costs. The extent to which ISM succeeds or fails in the future will depend in large part on the existence and utility of EMIS or an EMIS-like successor. During both the ISM PoP and ISM-X, EMIS was hampered by a series of problems that

---

[73]Information regarding plans and requirements for ISM is taken from Point Paper No. 8 prepared by Headquarters, U.S. Army Materiel Command, May 1, 1996, in response to the initial draft of this report.

[74]Current plans call for each COE to maintain a regional stock (ready-to-issue) of the ISM lines they repair.

interfered with maintenance operations, management, and billing. These types of problems are not unusual for prototype or developmental systems such as EMIS, and some or all of the problems may have been corrected in more-recent revisions of the software. More problems will undoubtedly arise until EMIS (or its successor) has become a mature, standard Army computer system.

In the draft cost-benefit analyses, no additional resources were budgeted for the future development of EMIS or its successor.[75] Such development is vital to the future of ISM, which will need a mature, fielded management system. The development process will not only have to correct any remaining problems in EMIS but also should enhance and expand the system. To reduce the data-entry workload observed during the ISM-X demonstration, EMIS should be redesigned to reduce the need for manual data entry. Allowing the EMIS database to be accessed simultaneously at multiple terminals would also simplify paperwork and information transfer at the installation.

Because data reliability has been a consistent problem during the ISM PoP and ISM-X, the future development of EMIS should also incorporate more-comprehensive validity checking to increase the reliability of maintenance information that it receives from legacy systems and manual input. Along these lines, maintenance units must also become more concerned and involved with improving the quality of data that EMIS receives from the legacy systems.

Finally, EMIS or its successor should have more connectivity with other supply, transportation, and resource-management data systems at the installation, regional, and national levels. This connectivity is needed to reduce associated workload of the installation staff. It would also provide the overall visibility of maintenance activities and inventories to the RSMM and NSMM that item managers, in particular, would need to make more-informed decisions about procurement. Furthermore, fielding schedules generally allocate the minimum-necessary version of EMIS to each type of participant at the local, regional, and national levels. Since improved information is one of the main benefits of ISM, we recommend that the Army consider fielding more-complete systems at all levels to allow wider sharing of EMIS information.

Time-Phasing of Benefits

During both the ISM PoP and ISM-X, EMIS and ISM fielding proceeded more slowly than planned, delaying the integration of new installations into the system. As a result,

---

[75]We understand, however, that Army headquarters has recently pledged $2 million annually for development of ISM automation.

ISM-X officially ran for only six months, rather than for the originally-planned nine. The Army found that installations cannot effectively participate in an ISM-type structure until EMIS has been installed and installation personnel have been adequately trained in its operation. Because some delays may also occur when an ISM-type structure is fielded as an operational system across CONUS, benefits should be estimated on the basis of a realistic schedule for participation by the new installations.

To be more specific, one of the draft cost-benefit analyses included an EMIS fielding schedule for CONUS that extended through the end of FY98. At the same time, however, it assumed that the Army would realize full CONUS-wide ISM benefits from reduced repair and requisition costs, starting in FY96. To improve estimates of ISM's economic benefits, repair-cost reductions and other benefits should be phased in no more rapidly than the fielding schedule, and might be more realistically assumed to lag behind the fielding schedule.

Differences Between ISM PoP, ISM-X Demonstration, and Implementation Plans

The ISM PoP and ISM-X differed from Armywide plans for ISM in ways that could influence cost and benefit projections based on the PoP and ISM-X. First, during the PoP and ISM-X, items were repaired and returned to the installation that sent them. Future implementation of ISM is intended to be based on repair and return to regional stocks held at the installation that does the repairs. Rather than wait for their own item to be repaired, customers will immediately receive a serviceable replacement from the stocks held at the COE installation. This arrangement should enable inventories of reparable components to be consolidated at the COEs.[76] To estimate the full benefits of ISM, the cost-benefit analysis should include estimates of those inventory effects that are based on the stocks the DBOF supply account at the COE would need to hold to cover demands while the COE was completing repairs on the unserviceable carcasses.

Repair-and-return-to-stock could also create a different set of financial incentives for customers and COEs than repair-and-return-to-owner. For example, under repair-and-return-to-owner, the retail stock fund at the customer's installation bears the risk that the item will be washed out and will need to be replaced through the wholesale system. However, under repair-and-return-to-stock, the AWCF DBOF supply account will bear the risk of washouts, which may change the behavior of COEs with regard to the number of

---

[76]Repair-and-return-to-stock could also improve readiness if an item was previously in short supply or if installations had insufficient funds to purchase replacements while their items were being repaired.

items they wash out or the behavior of customers with regard to the condition of the items they send to the COEs.[77] As a result, the future costs and benefits of ISM may differ from projections based on the ISM PoP and ISM-X.

Another potential change in ISM is the introduction of fixed-price bids on parts, rather than reimbursement for actual parts used. This could change bidding behavior, since COEs could sustain a loss on parts if they underbid. Or customers may send items back in worse condition if they do not have to pay for actual parts used. These changes in incentives could affect the future costs and benefits of ISM.

Finally, there is the issue of OMA versus Defense Business Operating Fund (DBOF) funding of Reparable Exchange (RX) stocks, which include ISM stocks. Currently, funding policies differ between FORSCOM and TRADOC, and they also differ at various installations within the two commands. These policies range from full OMA funding of RX stocks at Corps Support Commands within FORSCOM, to DBOF funding of stocks, parts, and some labor at some TRADOC installations. When RX stocks are held in OMA, ISM cost avoidance is retained in OMA dollars; however, when RX stocks are held in DBOF, part of the ISM cost avoidance may be retained in DBOF, unless RX credit rates are adjusted to reflect the lower costs to OMA customers.

As ISM is implemented, the Army intends to standardize these funding policies. RX stocks will be held in the FORSCOM and TRADOC retail stock funds, and installations will not be permitted to fund labor costs through DBOF. Thus, some changes in RX credit rates may be needed to ensure that ISM cost reductions are retained in OMA accounts, unless the MACOMs prefer to retain the cost avoidance in DBOF. Furthermore, moving some repair funding out of DBOF may limit the flexibility of some of the TRADOC installations that were using DBOF funds to reimburse labor costs. These changes may affect the potential costs and benefits of ISM.

Although some adjustments to the cost-benefit analysis could be made to account for these changes in stockage, bidding, and funding policies, it may be difficult to predict the behavioral responses of ISM participants as new policies are introduced. Therefore, it is important to continue to collect data on ISM performance as it evolves, both to ensure that the intended results are achieved and to refine estimates of the future economic benefits of ISM.

---

[77]Units can affect the condition of reparable components by removing piece parts or improperly storing, packing, or shipping items. Changes in the average condition of items sent to COEs could also affect the man-hours spent per item, and thus workload planning.

SUGGESTIONS FOR IMPROVING ISM COST-BENEFIT ANALYSIS

It is not an easy task to estimate the future economic costs and benefits of ISM. This section has discussed a number of recommendations that could improve existing estimates of the economic benefits.

First, the results of the ISM PoP and ISM-X have shown high variability in the return rates, productivity improvements, and cost avoidance achieved on each line. Therefore, we recommend that extrapolation to CONUS-wide benefits should be done on a line-by-line basis, rather than as a constant-percentage cost avoidance on all lines.

Second, an improved estimate of ISM economic benefits depends on comparing the costs incurred to repair an item in ISM with the costs that the Army would have incurred if the item had not been repaired in ISM. Therefore, it is important to ensure that price and cost factors reflect changes in costs at the installation and wholesale levels as realistically as possible. Retail prices do not reflect the avoided costs at the wholesale level when an item is repaired in ISM rather than purchased from the wholesale system, and so should not be used to estimate ISM benefits. Furthermore, fully burdened labor rates include costs that do not vary with maintenance activity. Therefore, we recommend that labor rates at both the installation and depot levels be adjusted to reflect the variable costs of repair as realistically as possible.

Third, since bidding strategies, stockage levels, and other forms of behavior may change as ISM evolves and as new policies are put into place, we recommend that data on ISM performance continue being monitored to ensure that the desired benefits are being achieved and to refine the estimates of future ISM economic benefits.

Finally, some additional measures could further improve the ISM cost-benefit analysis. In particular, the analysis could attempt to estimate the benefits of inventory reductions of reparable components and repair parts. Moreover, the analysis should consider the potential costs of additional personnel, equipment, and facilities, if actual productivity improvements do not meet expectations or if installations (under the pressure of competition) begin to duplicate facilities and capability existing elsewhere. To ensure that such duplication does not occur and that the ISM infrastructure is operating efficiently, both the RSMMs and NSMM should monitor repair operations and capacity at the installations. Lastly, the cost-benefit analysis should explicitly include the future costs of EMIS hardware and software development and fielding, and estimated benefits should not accrue faster than the system can be installed at the installations.

5. ISM AND WARTIME/CONTINGENCY OPERATIONS

This section addresses an issue that was not part of the ISM-X demonstration: how ISM would support a unit deployed during wartime or contingency operations. There are important differences between how ISM is evolving and the Army's current doctrine and practice for supporting deployed units. These differences have both resource and organizational implications. The Army should reconcile these differences now, while new logistics doctrine is being written. This section discusses the differences and suggests how they might be reconciled.

WHY LOOK AT WARTIME AND OTHER CONTINGENCY OPERATIONS?

We noted in our initial evaluation plan that the ISM-X demonstration was not set up to address wartime and other contingency operations and that we intended to consider them in our evaluation. Since ISM-X did not include wartime and other contingency operations, it is fair to ask why we include them in our evaluation, particularly since no data were collected during the demonstration. Wartime and other contingency operations may be full-scale military combat operations, such as the major regional contingencies (MRC) that form the basis for national defense planning, or they may be smaller, noncombat operations—operations other than war (OOTW)—such as humanitarian relief, peacekeeping, and disaster relief.

The Secretary of the Army has stated that "the Army's primary mission remains to fight and win the nation's wars."[78] A logistics system that works efficiently and cost-effectively in peacetime is important, but the wartime functioning of that system is more important. Since wartime and other contingency operations are so central to the Army's mission, we would be remiss if we did not identify and suggest changes to the relationship between ISM-X procedures and the Army's doctrine and practice for wartime and other contingency operations.

As the Army forges ahead with ISM, a comparison between the practices emerging from ISM-X and current Army doctrine and practice for supporting wartime and other contingency operations may yield important insights. Specifically, such a comparison will uncover any procedural conflicts, which are important to identify now for at least two reasons. First, the Army is in the process of revising its logistics doctrine, and any conflicts

---

[78]The Honorable Togo D. West, Jr., and General Gordon R. Sullivan, *A Statement on the Posture of the United States Army Fiscal Year 1996*, Washington, D.C., February 1995, p. ii.

should be resolved before that process is completed.[79] Second, differences may have organizational and resource implications that affect or are affected by other logistics initiatives, such as split-base operations, Velocity Management, and battlefield distribution. Early identification will allow the Army to address those implications as ISM is being implemented concurrently with other logistics initiatives.

ISM-X AND WARTIME/CONTINGENCY OPERATIONS DOCTRINE AND MAINTENANCE PRACTICE

Our analysis reveals some important differences between what occurred during ISM-X and Army doctrine and practice for wartime and other contingency operations. These differences have both organizational and resource implications. We take no position on how the Army should resolve these differences—whether by structuring the evolving ISM to match doctrine and practice for wartime and other contingency operations, the reverse, or some combination. We raise the issues here as something the Army should deal with before fully implementing ISM.

ISM and Army Wartime/Contingency Operations Doctrine

Two issues concern doctrine for deployed forces: one pertains to the role of the corps-level maintenance management organizations, and the other relates to the incorporation of business practices into doctrine.

Wartime/contingency operations doctrine. The doctrinal role for corps-level maintenance management organizations differs from that seen in ISM-X. Doctrine calls for an Army Service Component Command (ASCC) to manage sustainment maintenance (which would include general-support component repair) in-theater.[80] The ASCC did not figure into ISM-X. Logistics doctrine calls for divisions and corps to perform direct-support maintenance to repair a system; during wartime and other contingency operations, this maintenance level involves primarily replacement rather than repair of components. The unserviceable components that are removed from a system are then turned into the supply system, where a decision is made to either repair or discard them. The next step is to decide whether to repair the components in the direct-support unit or send them to general-support supply as

---

[79]Letter, 4th Corps Materiel Management Center, 13th Corps Support Command, AFVG-CMMC-CMMO, September 5, 1995, subject: "Split-Base Operations in a Corps Materiel Management Center," with enclosure letter, U.S. Army Combined Arms Support Command, ATCL-CG, February 26, 1994, subject: "Approval of Concept for Corps Materiel Management Center (CMMC) in Split-Base Operations." Headquarters, Department of the Army, *Logistics Support Element Tactics, Techniques, and Procedures*, Initial Draft, Field Manual 63-11, undated, circa 1995/1996.

[80]Headquarters Department of the Army, *Army Operational Support*, Field Manual 100-16, May 1995, pp. 4-12 through 4-17.

an unserviceable reparable for further inspection to make repair-or-discard decisions.
Logistics units above corps in the theater of operations have several roles: They provide
direct support to nondivisional and noncorps organizations assigned to their headquarters
and to other organizations under area-support arrangements; they provide back-up DS
support to divisions and corps; and they perform general-support maintenance for the supply
system.

The active-duty organization currently performing the general-support maintenance
management mission of the ASCC in overseas areas, to include wartime and contingency
operations, is the Theater Army Area Command (TAACOM). The Army is developing
doctrine that replaces the TAACOM with a Theater Support Command (TSC). The TSC
provides for flexible and modular organizations. However, for simplicity in this section, we
refer to the ASCC's maintenance management organization using the terminology of the
current force structure—TAACOM.

There are no active-duty TAACOMs in CONUS, although the force structure includes
several in the reserve components. (Appendix E provides background on the reserve
components and their participation in ISM.) Prior to ISM, installation DOL performed
general-support maintenance management in CONUS, except for the installations that have
an active-duty corps GS unit. In that case, the responsibility was split between them.
Although this general-support sustainment function has been a part of the corps structure in
the past, current Army doctrine does not assign GS maintenance units and management of
GS maintenance to the corps.[81]

These procedures differ from the two primary alternative courses of action developed
during ISM-X—called here "collaborative" and "centralized"—which have either the
peacetime corps or AMC operating the ISM regions in CONUS. By the end of the ISM
demonstration, the Army had not decided which course to follow. However, neither matches
current doctrine.

During ISM-X, some major commands and installations indicated that they favored
the collaborative approach, in which the Army owns the sustainment stocks but the
MACOMs control those issued to their commands, acting as "surrogate owners." After a COE
has repaired components, it might not return the exact components to the installation that
turned them in, but the same quantity of identical components returns to the controlling

[81]Field Manual 100-16, *Army Operational Support*, Headquarters Department of the Army, May
1995, pp. 2-10 through 2-14, 4-12 through 4-17. Field Manual 100-10, *Combat Service Support*,
Headquarters Department of the Army, October 3, 1995, pp. C-1 through C-4. Field Manual 63-3,
Corps Support Command, Headquarters, Department of the Army, September 3, 1993, Chapter 7.

installation's retail stock fund. Cross-leveling hinges on the cooperation of these "surrogate owners" and the financial arrangements that such transfers entail. New procedures may reduce the peacetime burden of keeping detailed financial records of transactions between installations within the same MACOM, but they may remain a few steps removed from the centralized management of sustainment stocks that is required to support wartime and contingency operations. The Standard Army Retail Supply System (SARSS-O) may provide the capability to cross-level within and between MACOMs, but the authority to direct such cross-leveling currently does not exist below HQDA.

Others advocated the centralized approach, which calls for AMC to operate both the NSMM and the RSMM. In this scheme, not only would AMC manage the workload allocation of reparables, it would also centrally control all sustainment assets. As reparables return to serviceable condition, they would be managed centrally and issued in accordance with Army priorities and guidance. The Army item manager might cross-level stocks rather than procure them to satisfy Armywide demands. Although this cross-leveling capability is possible under different management options, the centralized approach might require less coordination and therefore may result in more timely cross-leveling and in reduced pipeline requirements. A variant of the centralized approach was discussed by the Army. It would feature the Army National Guard and the U.S. Army Reserve managing their own reparable programs as separate activities, coordinating with the NSMM rather than with a corps RSMM.

Neither alternative squares with present doctrine. The collaborative approach has the corps performing tasks assigned elsewhere by doctrine. For example, in Europe and Korea there are functioning TAACOMs; in Kuwait, the Army Central Command manages maintenance through its Director of Logistics, Kuwait. If a corps or its elements were to deploy, it is not clear how its proposed ISM responsibilities for workload management would overlap or conflict with those of the theater army sustainment-maintenance organizations. A similar situation might arise during deployment of noncorps units, whose reparables are repaired in a corps-based region. These are not insurmountable situations, but they have not been thoroughly analyzed and planned for during ISM-X.

Furthermore, neither approach provides a role for the TAACOM. By Army doctrine, the ASCC organization manages the general-support maintenance for units deployed during wartime and other contingency operations. At one time, the corps had responsibility and the resources for this general-support sustainment function. In current and emerging doctrine, the ASCC has the responsibility, organizations, and equipment to perform the general-support maintenance management mission for the theater. In situations that do not warrant a large deployed force, the assets for managing and operating the GS maintenance mission may belong to other, subordinate Army commanders, but the function remains the responsibility of the ASCC.[82] In this light, the Army's ISM-X program may be seen as a good training opportunity for both reserve component TAACOMs and active-duty corps staffs, which might be called on to perform this role.

Business practices. A second doctrinal issue relates to the Army's incorporation of business practices in its logistics procedures. As the Army is revising its doctrine, it has the opportunity to incorporate various business practices with an eye to making its logistics operations simultaneously more efficient and less expensive. Emerging doctrine takes some of these practices adopted from industry into account. However, it will take a special effort to incorporate them all, because the traditional process for developing doctrine may not necessarily address business practices. Failure to incorporate the business practices into the new doctrine will result in wartime doctrine that differs from peacetime practice. The differences may make it more difficult for the new doctrine to support deploying units.

ISM and Current Maintenance Practice

Current maintenance practice for deployed units matches doctrine in using TAACOM resources. Both U.S. Army theater organizations in Europe and Korea (the 21st TAACOM and the 19th TAACOM, respectively) have responsibility for general-support maintenance, as does the DOL, Kuwait, an Army Central Command activity. Furthermore, a TAACOM, augmented by AMC resources and contractors, is providing general-support maintenance for the units now deployed to Bosnia.

Some of the peacetime TAACOM mission is performed by RC units, and war plans call for the RC to play a major role. The Army Reserve (USAR) has a relatively small peacetime repair mission, but it has a significant management-support mission for units outside CONUS. Reserves augment the 21st and 19th TAACOMs. The 21st TAACOM (CONUS Augmentation) has deployed personnel to Europe to assist the active Army organization in executing Operation Joint Endeavor. The 310th TAACOM Forward (USAR) provides continuing management for the Equipment Center Kaiserslautern, one of the sites in Europe that is supporting Operation Joint Endeavor. The 310th TAACOM also is providing staff augmentation to the 21st TAACOM at its peacetime location in Germany and at its forward location in Hungary. The 19th Corps Materiel Management Center has also been augmented in its forward location in Hungary by reserve components mobilized for deployment to

---

[82]Headquarters, Department of the Army, *Army Operational Support*, Field Manual 100-16, May 1995, pp. 2–10.

support Operation Joint Endeavor. The 19th TAACOM (CONUS Augmentation) has sent selected personnel to participate in RSOI (Reception, Staging, Onward Movement, and Integration) exercises in Korea with its active-duty counterpart. The 321st Theater Army Materiel Management Center (USAR) has been supporting training and contingency planning for U.S. Army Central Command operations in Kuwait since Operation Desert Shield.

As with doctrine, ISM-X procedures do not match current Army maintenance practices. ISM-X did not provide a role for the TAACOM; and although the reserve components participated in planning for installation maintenance and the allocation of EMIS hardware, their role in maintenance management for deployed forces did not figure into the ISM-X demonstration.

IMPLICATIONS OF DIFFERENCES IN DOCTRINE AND PRACTICE

These differences in doctrine and practice have important organizational and resource implications. From an organizational perspective, the corps currently do not have the connectivity, organization, or staffing to fulfill the ISM role for a deployed unit. Once a unit moves to an overseas theater, the corps may lose visibility of supply and maintenance transactions, a critical aspect of ISM procedures. Furthermore, their organizational structure does not authorize the corps the necessary material to pack the repaired components nor the personnel or transportation to move them during wartime and other contingency operations.

Turning to resources, practices emerging from ISM-X may affect both stockage levels and forecasting. Under current practice, units mostly repair components at the retail maintenance activity in peacetime and mostly replace them through the wholesale supply system in wartime. Practices emerging from ISM-X tend to exacerbate this tendency by fostering retail repair over wholesale replacement, causing the demand for reparable components to fall while increasing the demand for spare parts.

This difference affects both the type and level of stocks shipped to a theater to support a deployed unit and the war reserve stocks. Wholesale-supply demand feeds into the planning process that establishes the production and procurement programs. Because wholesale demand reflects the peacetime preference for repair, the production and procurement programs will follow suit. Thus, both the contingency planning and funding processes may overestimate the need for repair parts to fix components and underestimate the need for replacement components.

SUGGESTIONS TO FACILITATE ALIGNMENT BETWEEN PEACETIME, WARTIME, AND OTHER CONTINGENCY OPERATIONS

Operation Desert Shield provided the primary impetus for justifying the initiation of the ISM concept as an Army program. However, the expectations of cost savings and avoidance diverted attention away from wartime and other contingency operations and toward peacetime garrison efficiencies. Unless wartime and other contingency operations are considered in ISM implementation, the Army may again find itself with a logistics system that complicates rather than improves support to deployed forces.

The most critical aspects for successful ISM performance—visibility, maintenance processes, and retrograde of unserviceable reparables—take on a different hue in deployment (war or operations other than war) from those in garrison. Deployment operations to wartime and other contingency operations areas must be considered as a claimant on resources (e.g., people, equipment, funding). Therefore, the reallocation of these resources needs to be examined in a trade-off analysis between projected wartime and other contingency operations requirements and projected ISM cost savings and avoidance for garrison operation.

The Army should take maximum advantage of this opportunity with ISM to align the requirements, funding, and maintenance and supply processes for both garrison business practices and its wartime or deployment missions.

We suggest a few changes that will facilitate the necessary alignment:

- A first step to bringing forecasts in line with demand involves providing the retail supply and maintenance demand as an input to both the wartime/contingency-planning processes and the Planning, Programming, Budgeting, and Execution System (PPBES). By incorporating the ISM component-repair quantities as a portion of retail-supply demand, the overall retail demand for reparable components will more closely approximate the behavior expected in wartime and other contingency operations—replacing components instead of repairing them. This broader demand may result in more-appropriate budgets, more-available stocks, and a forecast that more closely approximates actual demand.

- The Army's Velocity Management initiative is exploring methods to improve logistics processes in peacetime garrison operations and when deployed, e.g., order-ship, repair cycle, and transportation. During wartime and other contingency operations, Velocity Management will operate in conjunction with the Army's Battlefield Distribution System program (BDS), which has suggested innovative methods to improve transportation and supply support in wartime and other contingency operations areas, whether in an MRC or an OOTW. The anticipated disparity in component demand and repair-parts demand under current doctrine might have been accommodated by the large stockage quantities held at various organizations under prior logistics practices (the "iron mountains"). Under the new programs, much less stock is intended to be held at all locations. The intent is to use a combination of supply and transportation policies resulting in a distribution system that is more timely and accurate. The discrepancy between component demand and repair-parts demand might be ameliorated at the customer end by improvements suggested by these two Army initiatives. Bringing the disparate demand together at the national-level stockage position will necessitate a close linkage between item managers, COE managers, and in-theater wartime/contingency operations managers.

- The Army is developing a Theater Support Command to support wartime and other contingency operations. The NSMM could facilitate the linkage between COEs, corps regional maintenance managers, the theater maintenance manager, and the wholesale item managers if the he/she were represented in the AMC–Logistics Support Element that deploys with the TSC.

- The emerging TSC will be built heavily around the existing TAACOMs. The emerging TSCs should be given the hardware, software, and training to perform their theater sustainment-maintenance-management mission as part of the ISM program. Personnel could then be trained during exercises, such as Prairie Warrior (Force XXI), Atlantic Resolve (ACOM and EUCOM), Vigilant Warrior (CENTCOM), RSOI and Ulchi Focus (Korea), and Yama Sakura and Cobra Gold (U.S. Army Japan). Reserve component members of the TSC could be detailed to work with the peacetime regional sustainment-maintenance managers during their individual and unit training periods to learn the techniques and acquire the requisite management expertise.

- At the end of the ISM-X demonstration, a doctrinal concept was being prepared that places the ISM regional sustainment-maintenance management in the corps rather than in the Theater Army Support Command. However, the necessary organizational changes have not yet been specified (e.g., the quantity and specialty of personnel, the necessary electronic information systems and communications connectivity, and the identification of packaging or other special-purpose equipment). The organizational changes must be identified, funds must be allocated, and personnel must be trained early to enable corps regional sustainment-maintenance management to operate in wartime and other contingency operations.

6. POLICY CHANGES TO ENHANCE ISM

In this section we first discuss policy issues raised by ISM, including the effect of changes in the depth of repair, the selection of lines for ISM repair, the effect of personnel and pay constraints, financial constraints on the COE bidding process, and the discrepancy between local OMA cost avoidance and the Army's cost avoidance from ISM. Although these issues arose in the context of the ISM PoP and ISM-X, they are influenced by policies outside the scope of ISM itself. Next, we suggest some policy changes to address these issues and to enhance the operation of ISM and its benefits to the Army. Our recommendations include serial-number tracking and level-of-repair analysis to determine the most cost-effective repair standards for each line; changes in credit policy to give OMA-funded organizations financial incentives to repair the most cost-effective lines from the perspective of the Army as a whole; and funding ISM through the stock fund to allow greater labor flexibility and more-realistic COE bids.

POLICY ISSUES RAISED BY ISM

Changes in the Depth of Repair

Under ISM, a short-term cost avoidance results from repairing more lines to an IRON standard at the COEs, rather than purchasing new or overhauled lines from the wholesale system. However, little is known about the long-term implications, for both readiness and total costs, of changing the depth of repair. If new items are not purchased as frequently, the average age and service life of the existing stock will increase. Fewer lines will have depot-level repairs, which include more-extensive replacement of subcomponents. Higher breakdown and failure rates, and more-frequent repairs over time, could be the consequence. Conversely, in the past, perhaps lines were being repaired too extensively or replaced too quickly, before the end of their useful service life. Thus, reducing the depth of repair could result in cost avoidance without impairing readiness.

Currently, there are no data to support either contention. We recommend below that serial-number tracking or some other approach be used to gather information about the cost-effectiveness of changes in the depth of repair and to adjust the periods of service between specific maintenance actions.

Selection of Lines for ISM Repair

During the ISM PoP and ISM-X, COE lines were selected from lines that were on at least one installation's RX list. The selection process used specific criteria, including the following:[83]

1. The line must support a major weapon system or be a significant readiness line.

2. Total cost to repair (including transportation and packing) must be less than 80 percent of the net cost to purchase.

3. End-item demands are common to at least two installations, and there are at least nine total demands per year (ground reparables) or three total demands per year (aviation reparables).

4. The line must currently be repaired by one installation with at least six repairs per year (ground reparables) or three repairs per year (aviation reparables).

5. The washout rate must be less than 50 percent.

Potential readiness effects were determined from asset posture and the availability of items through the wholesale system. Finally, the selection priority was assigned on the basis of potential cost avoidance per labor hour.[84] During the ISM PoP and ISM-X, lines were selected largely to obtain the highest potential OMA cost avoidance for the installations. Therefore, as ISM expands, fewer of these desirable lines will likely remain, and additional cost avoidance per item may be lower.

When considering potential RX lines to add to ISM in the future, the RSMMs should carefully assess lines' suitability for regional repair. For a number of reasons, regional repair operations may not be the best option for all candidate components; the distribution of repair cost avoidance shown in Tables 4.5 through 4.8 provides evidence to support this contention. Lines with very low demand rates and equipment density may not generate enough demands to realize any improvement in repair efficiency at the regional level. These lines might be repaired more cost-effectively at a single, national location, such as a depot. Lines with high demand rates at multiple installations may be more cost-effectively repaired at the installations where the demands occur. The same may be true for other lines that are difficult to pack or ship or that are particularly easy and inexpensive to repair.

Before adding a line to ISM or estimating its potential benefits, the Army should carefully consider whether regional repair will be the most effective approach to ensure

---

[83]See Computer Systems Development Corporation (1994) for a more detailed discussion.

[84]*Cost avoidance per labor hour* was calculated for each bid on spreadsheets distributed at PP&C 6. It is defined as expected cost avoidance per item divided by the number of labor hours bid per item.

maximum readiness at minimum cost with acceptable response times for that particular component. We recommend that the RSMMs and the NSMM consider the Army's overall repair-capacity needs when selecting COE lines and use level-of-repair analysis over the lifetime of component use to adjust levels of repair when necessary.

Personnel and Pay Constraints

As repairs move from less-efficient to more-efficient GS maintenance operations under ISM, the personnel demands in these repair activities will change accordingly. Installations that lose repair activities should retain fewer direct-labor personnel. Installations that gain COEs because of their higher productivity and lower costs should be allowed to grow if necessary. ISM will achieve optimal efficiency in the long run only if personnel slots can be shifted easily among installations.[85]

Current policies, however, make such a process difficult or impossible. During the ISM PoP and ISM-X, the direct-labor costs of repairs were funded by OMA appropriations and could not be reimbursed between installations (except for national work and repairs performed by and for National Guard units). The civilian-pay cap imposed on nonreimbursable work by FORSCOM and the personnel caps imposed in general by the Army do not allow installations that win COEs to increase their civilian workforces. Moreover, because installation budgets have been tightly constrained, DOLs that contract for their maintenance operations cannot easily obtain the additional resources needed to expand their maintenance contracts. If their workload rises and productivity gains do not sufficiently expand their capacity, they must seek permission to increase their maintenance contracts and request budget supplements from their MACOM or the Army to pay for the increased workload. As long as ISM remains small, this constraint should not be a problem. However, it will become increasingly important as ISM expands.

On the other side of the problem, if depots or installations lose significant amounts of maintenance work under ISM, both their workforces and repair facilities should shrink. These organizations could use the excess capacity to attack backlogs, but in the long run they will have to reduce their capacity. They will understandably be reluctant to reduce their direct labor and indirect personnel. In the current environment, these organizations realize that they cannot easily regain lost workforce if conditions should change. Moreover, installations understand that losing positions (and skilled labor) will inhibit or prevent them from effectively bidding for new COE work or rebidding COE work.

---

[85]Shifting tools and equipment from less-efficient to more-efficient maintenance facilities may also enhance the benefits of ISM.

These related problems have several implications. First, the COE workforces can shrink (albeit slowly) but cannot easily expand. Thus, once the maximum capacity of the more-efficient organizations has been reached, transfer of maintenance work under ISM from less-efficient to more-efficient organizations will halt or at least be severely limited.[86] Moreover, in time, this situation will also tend to reduce the Army's full-time repair-and-maintenance workforce. Since this smaller workforce might be sized to cover normal peacetime operations or OOTW, it might need to be augmented by overtime work or reservists during major regional contingencies and other surges in activity. Any proposals for reduced capability should therefore be studied in a trade-off analysis to address maintenance stockage policies for peace, OOTW and MRC operations; close coordination with other logistics initiatives (e.g., Velocity Management, Battlefield Distribution System, Split-Based operations) being concurrently implemented may ameliorate the adverse effects of these reductions.

The Army should address these issues if it wants to achieve the maximum benefit from the ISM maintenance concept. As we recommend below, ISM repairs should be stock-funded so that ISM labor would be reimbursable; thus, civilian pay caps that apply only to nonreimbursable labor would not apply. This approach would allow maintenance workforces to expand at the most efficient installations and shrink elsewhere.

The COE-Selection Process

During the ISM PoP and ISM-X, COE bids were only loosely tied to payments for repairs. Bids were evaluated on the basis of parts costs and man-hours at fully burdened labor rates, but COEs were reimbursed only for transportation, packaging, and actual parts used.[87] Because COE labor is OMA-funded, it is not reimbursable between installations, nor do any funds change hands to cover indirect and G&A costs. To link bids and actual costs, COE lines were rebid if actual costs per repair exceeded the bid by more than 25 percent. However, this rule still allows a great deal of discrepancy between bids and performance. As a result, to protect their labor forces, installations may be tempted to try to win COE lines by entering bids that are lower than their true expected repair costs. Although the ISM-X data

---

[86]A maintenance organization's maximum capacity is based on current resource-allocation patterns, including both personnel and physical facilities. Although personnel constraints might best be alleviated by shifting personnel between organizations, it is likely to be more cost-effective to alleviate capital constraints by transferring better management techniques to underutilized facilities than by building new facilities.

[87]Labor was reimbursed for National Guard, National Training Center (NTC), and national work. Future plans for ISM call for fixed-price bidding for parts, which would still allow installations to underbid on labor costs.

indicate that, on average, bids were not below historical repair costs, at PP&Cs 6 and 7, there appeared to be underbidding on a few lines by some installations.[88] As ISM spreads to more installations, more-intense competition may result in further underbidding. Therefore, projected cost avoidance based on COE bids may overestimate actual cost avoidance.

In addition, a GS maintenance facility could continue to win COE lines even if total expected repairs exceeded the facility's repair capacity. During ISM-X, a Surge Management program was instituted to shift workload to other installations if a facility was unable to keep up with its production schedule because of capacity constraints. However, if COE lines are not consolidated at a single installation, the full benefits of maintenance efficiency and inventory reduction may not be achieved.

In general, as long as bids and payments are not tied to actual variable repair costs, customers and COEs are likely to continue to have incentives to "game the system," e.g., COEs may submit bids that do not reflect their true repair costs. For example, if there is fixed-price bidding only on parts, installations may still underbid labor hours to win COEs. Customers may eventually begin to reduce their direct OMA expenditures on DOL maintenance workforces, since they can get ISM repairs by paying only for parts. If ISM repairs are stock-funded, as we recommend below, then all ISM repair costs would be reimbursable between installations, and full fixed-price bidding (with payments equal to bids) could be implemented. This change could reduce some of the incentive problems associated with the ISM PoP and ISM-X bidding process and lead to more-efficient operations in the long run.

Disparity Between Local and Army Cost Avoidance

Another issue discussed in Section 4 in relation to cost-benefit analysis is the difference between local OMA cost avoidance and the Army's cost avoidance when items are repaired at the installation. Local and Armywide financial incentives could be aligned if prices charged to OMA customers reflected the Army's cost avoidance. In other words, price and credit policy could be thought of as a signal to installation-level decisionmakers about whether to do repairs themselves or to send them to the wholesale system.

Under current policy, there is a gap between what OMA customers pay and the Army's cost avoidance, because retail prices include a Supply Management surcharge to cover supply-system overheads, and because repair costs are averaged across a wide variety of

---

[88]Lionel A. Galway, "ISM-X Evaluation, Individual Item Repair Performance," unpublished RAND research.

items. Thus, from the perspective of the Army as a whole, ISM may not be targeting the most cost-effective lines for repair.

Currently, the prices that OMA customers pay for RX repairs (of which ISM repairs are a subset) do not reflect the costs of repair either. RX credit policy varies somewhat by installation, but generally covers only washout costs at FORSCOM installations and may also cover the costs of parts and some labor at TRADOC installations.[89] Other RX repair costs are directly funded by OMA appropriations to GS maintenance activities. Thus, in general, the prices of RX repairs are lower than the variable costs of repair, which is likely to cause OMA customers to prefer RX programs to other sources of repair, particularly to the wholesale system.

We recommend that credits for Army-managed items reflect the wholesale cost of repairing or replacing the items, so that installations would have a financial incentive to repair items only if it is cost-effective for the Army as a whole. Furthermore, if ISM repairs are stock-funded, then payments by OMA customers can be set equal to winning COE bids, which should reflect the actual cost of ISM repairs.

SUGGESTIONS FOR IMPROVING ISM

To address these issues, and to enhance the cost-effectiveness of ISM, we recommend three policy changes:

- Use serial-number tracking and level-of-repair analysis to determine the most efficient depth of repair for major lines.
- Set retail prices paid by OMA customers equal to the variable costs of repair at the installation (RX) and wholesale levels so that customers can make appropriate repair decisions.
- Fund ISM repair costs through the retail stock funds[90] so that all variable costs (labor, parts, transportation, and washout costs) can be reimbursed.

We recognize that these changes go beyond the scope of ISM and will require further groundwork before they can be implemented. However, we believe that it is important for the Army to consider these changes now in order to enhance the performance of ISM. We discuss each of these recommendations in more detail below.

---

[89]Army policy has recently been changed to preclude DBOF funding of RX labor costs after the end of FY96.

[90]This recommendation can also be implemented when the Army makes the transition to a single stock fund, provided that separate balances are maintained for installation-level ISM programs.

Using Serial-Number Tracking

Serial-number tracking could be used to assess the cost-effectiveness of different repair standards by linking successive repair episodes for an individual reparable component and comparing the time between repairs for components subjected to different frequencies and depths of repair. However, recording serial numbers on components under repair can be time-consuming for maintenance personnel, and the analysis of the serial-number data can be complicated.[91] It may therefore be most appropriate to focus on a small number of high-value components.

Such tracking can have additional benefits, such as identifying "bad actors," i.e., the small proportion of defective components that cause a large proportion of removals and replacements.[92] It is often better to dispose of these defective components than to continue repairing them.

Reviewing Level-of-Repair Analysis

The RSMMs and the NSMM can use level-of-repair analysis, combined with information on the frequency of demands and the capacity of repair equipment, to determine more accurately whether repairs should occur locally or should be consolidated to the regional or national level. Under current policy, level-of-repair analysis routinely occurs only when a line first enters service. Over time, technologies, capabilities, or the amount of equipment in service can change, leading to adjustments in the appropriate level of repair. Thus, it may be useful to repeat level-of-repair analysis over the lifetime of a reparable component (or the associated end item) to make these adjustments.

Setting Retail Prices Equal to Repair Costs

Currently, stock-fund prices are set to ensure that the stock funds break even. However, this pricing policy sometimes creates incentives that work against the best interests of the Army, as we have illustrated for some of the ISM-X lines. To align the incentives of local decisionmakers with the Army's objectives, we recommend that both RX

---

[91]Army maintenance information systems record the serial numbers of repair parts installed in a reparable component. While there is disagreement about the quality of these data, we have observed that, in some cases, the serial numbers appear to be recorded consistently and accurately. However, although the serial number of the reparable component itself can be recorded on maintenance paperwork, there is no designated space for this number, and it is rarely found in maintenance databases.

[92]See, for example, J. R. Gebman, D. W. McIver, and H. L. Shulman, *A New View of Weapon System Reliability and Maintainability: Executive Summary,* Santa Monica, CA: RAND, R-3604/1-AF, January 1989.

and DLR credit rates reflect the variable costs of repairs at the installation and wholesale levels, respectively.[93]

If variable costs are reflected, local ISM cost avoidance would match the Army's ISM cost avoidance, so OMA-funded units would have the right incentives to make decisions that benefit the Army as a whole.[94] It would also help ISM to target the lines for which it has the greatest cost advantage, without collecting additional information from the wholesale system. For some lines, OMA customers could also be allowed to choose whether to buy an IRON-repaired item from a regional ISM COE or an overhauled item from the wholesale system.

If retail prices paid by OMA customers were equal to the wholesale variable cost of repairs, the wholesale system would need some other mechanism to recover its overhead costs. One possibility might be direct OMA funding of overhead costs, or lump-sum payments from OMA customers (such as divisions or MACOMs), proportional to their use of the wholesale logistics system.[95] These options would need to be more thoroughly analyzed in cooperation with the Army's Financial Management community to determine their feasibility.

Funding ISM Repairs Through the Stock Fund

Our third recommendation, involving OMA as opposed to DBOF funding, requires a more detailed explanation than the other two. Figure 6.1 shows the current funding patterns for ISM and other RX lines at most installations. The financial policy for ISM seems to be moving in the direction of using DBOF (currently the Retail Stock Fund) to fund RX stocks, but paying for ISM parts and labor with direct OMA funding of GS maintenance.[96] OMA customers would pay only the expected washout costs. Thus, if RX credit rates are set too

---

[93]This policy would apply to items for which the Army wants local decisionmakers to choose the most cost-effective sources of repair. For some items, the Army may have other reasons (e.g., mobility) to prefer that repairs take place at a particular level. In these cases, prices could be used to give OMA customers the incentive to choose the correct level of repair, i.e., by making the correct repair source the least expensive one. For further discussion, see Ellen M. Pint, "Economic Incentives in the Use of Army Stock Funding," unpublished RAND research.

[94]Similar recommendations have been made for Air Force stock-fund pricing policies. See William P. Rogerson, *On the Use of Transfer Prices Within DoD: The Case of Repair and Maintenance of Depot-Level Reparables by the Air Force,* McLean, VA: Logistics Management Institute, PA303RD1, March 1995.

[95]Currently, some stock-funded activities use similar pricing policies. For example, Air Force Air Mobility Command sets its prices to be competitive with commercial carriers, and covers the costs of its surge capacity with direct OMA appropriations.

[96]These plans require COSCOM RX stocks to be capitalized into DBOF and OMA budget holders to be compensated, and some parts and labor that were funded through DBOF (by TRADOC, for example) to be funded by OMA.

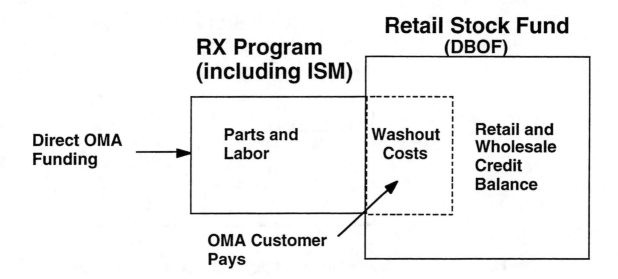

Figure 6.1—Current ISM Repair-Cost Funding Pattern

low, some ISM cost reductions will be retained in DBOF. However, it is difficult to determine whether RX credits balance washout costs, because the balance is maintained only for the Retail Stock Fund as a whole. As a result, any net profit or loss on an RX program tends to be swamped by net profits or losses on the balance between wholesale and retail credits, which involves a much larger number of transactions. Therefore, ISM cost reductions may simply be subsidizing losses on the wholesale/retail credit balance, rather than being passed through to OMA customers.

To avoid these problems, we recommend that all the variable costs of RX repair (including ISM repair) be moved inside DBOF, but that a separate balance be maintained for RX programs. (See Figure 6.2.) RX credit rates for OMA customers would then reflect the variable costs of repairs, including parts, labor, and washout costs. Because a separate balance would be maintained, ISM cost reductions could be passed through to OMA customers by adjusting credit rates, rather than potentially being absorbed by losses on the wholesale/retail credit balance.

DBOF funding of RX programs would also address some of the other financial issues associated with ISM.

First, OMA customers could pay for ISM labor, which would allow fixed-price bidding for regional ISM COEs, including parts, labor, transportation, and washout costs. If RX programs were required to maintain a separate balance within DBOF, COE bids would have to be realistic in order to cover expected repair costs. Maintenance facilities that underbid would have an incentive to offer money-losing lines for rebidding, because they would have to

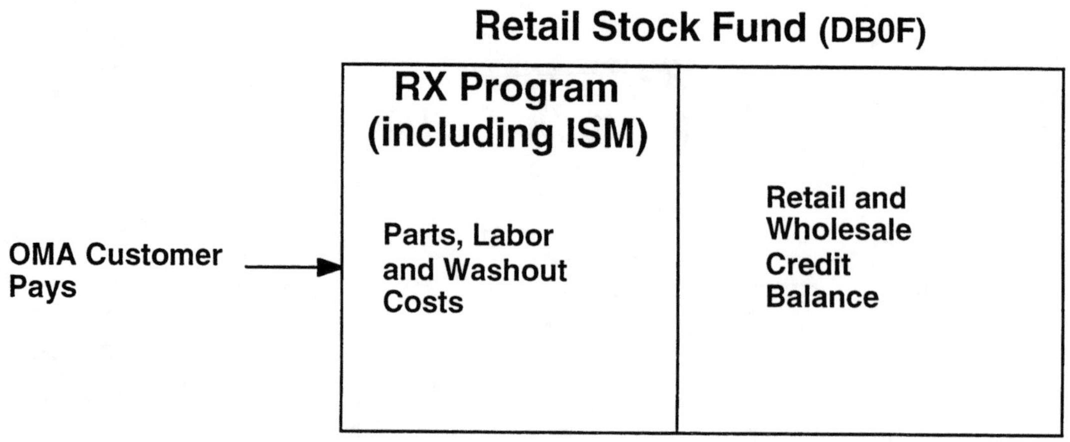

Retail Stock Fund (DB0F)

RX Program (including ISM) Parts, Labor and Washout Costs	Retail and Wholesale Credit Balance

OMA Customer Pays →

Figure 6.2—ISM Repair-Cost Funding Within the Retail Stock Fund

cover losses with their own OMA funds. However, if COEs were earning profits on some lines, RSMMs might need to initiate rebidding.[97] It may be necessary to monitor "profits" and "losses" on individual lines to ensure that cross-subsidization does not occur.

Second, reimbursable labor would allow organic and contractor maintenance facilities to adjust their workforces to reflect changes in their COE lines. As a result, efficient facilities could grow, but inefficient facilities would have to shrink. However, it would not be appropriate to use fully burdened labor rates to bill for ISM labor. Labor rates should be adjusted to reflect variable maintenance costs that are currently being directly funded with OMA dollars.[98]

As in the case of setting retail prices equal to wholesale repair costs, DBOF funding of ISM repair costs would require further analysis to determine an implementation strategy. The analysis would have to resolve distinctions between OMA-funded and DBOF-funded activities at each installation DOL, for example. Nonetheless, we believe that it is important to the efficient functioning of ISM in the long term for the Army to take these issues into consideration.

---

[97]Rebidding rules should be carefully determined to avoid eliminating the incentives of COEs to reduce costs below their bids. If COEs are allowed to keep the gains from additional productivity improvements for a fixed period of time (one to two years, for example) before rebidding occurs, these incentives could be preserved while still ensuring that the gains will eventually accrue to customers when rebidding occurs.

[98]Current Army efforts to implement Activity Based Costing, a form of cost accounting that identifies the variable costs of individual activities, can help to identify costs that vary with maintenance activity.

7. CONCLUSIONS

The ISM-X demonstration has had a number of positive results. The Army has more visibility than before of maintenance capability and capacity. This visibility is critical to effective management of maintenance personnel and facilities. Furthermore, the information collected during ISM-X indicates some improvement in maintenance processes at the COEs, e.g., the median man-hours to repair ISM-X parts has decreased from that of pre-ISM repairs. Moreover, repair across MACOMs was successful. At the beginning of the demonstration, no one was sure how well sharing maintenance responsibilities across MACOM lines would work; however, the sharing worked so well that we now must remind ourselves that it was once an area of uncertainty.

OVERVIEW OF RAND's CONCERNS AS ISM IS EXPANDED ARMYWIDE

RAND recognizes that the ISM PoP and ISM-X demonstrations were tests of integrated sustainment maintenance. And because of many circumstances, some beyond the Army's control, parts of the ISM concept were not tested or had to be modified. In the preceding sections we highlighted areas of the ISM-X demonstration that were beneficial to the Army, such as new visibility of maintenance capacity and capability. In this section we highlight specific topics that could negatively affect ISM as it is expanded Armywide:

- As financial resources diminish, the Army is looking for ways to save money. Cutting budgets based on projected ISM cost savings may not be wise. Our research shows that the Army must be extremely careful in reducing budgets based on cost-avoidance projections from the ISM PoP or ISM-X. One of the reasons for this caution is that changing financial policies create different economic incentives for various Army organizations and thus may encourage different financial decisions.[99] Additionally, the data on which to base economic decisions lack the quality necessary for ensuring correct actions.

- Long-term cost savings can come only from reducing both procurement and infrastructure. In a short demonstration such as ISM-X, it is not reasonable to trim the workforce; ISM will achieve optimal efficiency in the long run only if

---

[99]For example, the implementation of a single stock fund has the potential for changing perceived savings at the local level. Also, very different financial incentives operate under repair-return-to-sender (as was practiced during the tests) compared with repair-and-return-to-stock (as may occur when ISM expands).

personnel slots can be shifted easily among installations and decreased in response to decreased demand and increased production. Additionally, the Army must be cautious about building new facilities that duplicate existing ones. Because money for facilities comes from different budgetary sources than money for logistics, there is a tendency to continue building facilities that have been previously approved but may no longer be required.

- The laudable emphasis in ISM PoP and ISM-X on cost avoidance tended to overshadow the importance of organizational structure and authority for achieving ISM's overall goals. The two tests lacked emphasis on the hierarchy of responsibility as discussed in Section 2.

- During ISM-X, the NSMM should have assisted the MSC item managers in passing "national work" down to the LSMMs. In fact, the NSMM was usually bypassed during the negotiations and awarding of national work. Additionally, partly because the information systems used by the item managers are not compatible with the Executive Management Information System and because all dealings with the NSMM, RSMM, or LSMMs had to be documented manually, the NSMM could not work with the item managers to identify wholesale requirements, review repair-or-buy decisions, or effectively maximize cost avoidance or reduce procurement at the national level. While the lack of adequate communication was a clear problem in ensuring the visibility required to fulfill the initial expectations for ISM at the national level, the primary problem was the failure to integrate the NSMM into ISM and provide the position with the authority to accomplish its goals.

- The Army Reserve Component participated only in ISM-X planning, not production. The various National Guard units that did participate were willing to participate much more than they were tasked. If this situation continues, it will be difficult or impossible to develop regional plans for reparable programs and training requirements with Reserve Component units. Under the original ISM concept, the RSMM would have integrated these plans to develop regional training plans for all units. The training benefits of ISM to the USAR were not fully examined.

- During the ISM-X demonstration, actual returns of unserviceable carcasses fell below planned returns. This shortfall in unserviceable returns is an important problem for the Army as a whole and is critical to the resource planning that is one of the primary goals of the ISM concept.

- Under ISM-X, items were repaired to an inspect-and-repair-as-necessary standard at the COEs, rather than purchasing new or overhauled items from the wholesale system. The long-term implications of changing repair practices on both readiness and total costs is not known. It is possible that this change has differential effects, being cost-effective for some lines and costly for other lines.

- ISM continues to evolve even as this report is written. Prior to the conclusion of the ISM-X demonstration, the planning for ISM had not addressed contingencies. However, the Corporate Board has since requested that the United States Army, Combined Arms Support Command (CASCOM) initiate development of a doctrinal concept to reflect the emphasis of ISM. CASCOM is now developing that concept. There is a concern that deficiencies noted in ODS might not be resolved with current implementation plans.[100] The Army has an opportunity to resolve deficiencies noted in ODS if it brings evolving logistics doctrine and current implementation plans into a closer alignment, as well as addressing those concerns in conjunction with implementation of other logistics initiatives (e.g., Velocity Management, Battlefield Distribution system, and the Theater Support Command Structure). Some suggested alignments are discussed later in this section.

SUGGESTIONS FOR IMPROVING ISM

To realize the full benefits of ISM, the Army will need to undergo some changes. This subsection discusses the areas in which we believe changes are required and develops the rationale for them.

COE-Selection Process Requires Improvement

It is critical that the RSMM monitor the performance of the COEs. Elapsed time, man-hours, parts costs, and depth and quality of repair are all indicators of a COE's performance. To let the COEs see how their performance compares with that of others making similar repairs, these measures should be routinely collected and made available to all the installations in the RSMM's region.

Such monitoring will also help the RSMM identify which COEs need to be rebid. Rules for rebidding of COEs must be established and agreed on by all participants. Careful

---

[100]The seventh Production, Planning and Control meeting (PP&C 7) was held at Fort Bliss, March 19–20, 1996. At this meeting, ISM repair of TAACOM items for the Bosnia deployment was discussed.

rebid rules and monitoring of performance will help prevent monopolistic practices from developing in ISM.

We suggest that the Army consider making COE repairs reimbursable work and hold COEs to their bid prices—meaning that COEs can charge parts and labor only up to the bid price. To make COE repairs reimbursable work will require changes in the financial system as discussed in Section 6.

Item Managers at the MSCs Must Be Integrated with the NSMM

The management of the Heavy Expanded Mobility Tactical Truck (HEMTT) engine illustrates why item managers need to be integrated with the NSMM. Many HEMTT engines and transmissions were purchased during ODS. After the contingency, they were considered to be in excess by the wholesale supply system, so units did not get credit for returning "excess." HEMTTs continued to be used in the field for exercises, became unserviceable, and required more repair than was usually done in the DOL. As a result, the wholesale and retail systems got out of balance. Exacerbating the problem, the wholesale source of repair disappeared as depots closed. A shortage of HEMTT engines went unnoticed and the lack of serviceable HEMTT engines was degrading readiness, so TACOM was going to need to make an emergency buy. Fortunately, TACOM could use ISM-X COEs as a source of repair, and received repaired engines and transmissions from them much faster than it could replenish its stock through procurement. TACOM's item manager arranged for these repairs of ISM items without NSMM participation.

This story illustrates the importance of visibility of wholesale and retail stocks, repair capability, and capacity. It is also an example of how some current Army processes impede sensible logistics practices. The integration of the item managers with the NSMM would help prevent a repetition of the HEMTT engine problem.

The NSMM must also be given authority to redistribute specialized repair equipment and maintenance workload, since the full maintenance capability and capacity across the Army will be visible only to the NSMM once ISM is fully implemented.

Quality-of-Maintenance Data Need Improvement

The Army maintenance data collected in EMIS for the RSMM's use in managing and analyzing ISM comes primarily from Army legacy systems. Section 3 presented some of the data problems encountered during the analysis of ISM-X performance, including missing parts costs or man-hours, gaps in dates in the EVAC file, and missing action dates in some work orders. Missing information is of particular concern, because the subject line items were intensely managed and scrutinized, and the only money that changed hands during the ISM-X demonstration was for parts.[101]

As ISM encompasses more critical lines and as more-sophisticated tools are used to manage the program (such as the new requirements-determination and decision-support modules in EMIS), accurate data will be ever more important. And there will not be sufficient manpower to engage in detailed manual review and correction. Consequently, three steps need to be taken:

1. The LSMM/COE staffs need to take primary responsibility for data quality, both reviewing and checking it locally, and emphasizing to local personnel how critical good data are in evaluating their own performance. Accurate data are as important to ISM in the long run as is quality repair.[102]

2. The RSMM should continue to evaluate data quality coming from each COE and work even more aggressively with the COEs to improve their data (possibly rejecting bad data).

3. The ISM management should work toward maintaining a long-term database on ISM work for analysis, with errors corrected as they are detected.

Users of ISM Information Must Distinguish Between EMIS Problems and Data Problems

EMIS is a front end for standard Army data systems; if problems exist with Army data, problems exist with information from EMIS. However, we did see problems with EMIS not pulling labor hours data from the Maintenance Information Management System (MIMS).[103] Thus, the problems with information from EMIS may be compounded ones.

Having visited all the EMIS sites in the ISM-X demonstration, we know that the users have many suggestions for changes and improvements to EMIS. Most of these changes or improvements are fundamental requirements rather than optional frills. EMIS, like ISM, is evolving and improving. During the demonstration, new versions of EMIS came on-line to

---

[101]All other costs of repair were borne by the COE, except for transportation costs, which were paid by FORSCOM if the shipment was by Federal Express or by the sending installation if shipped by another carrier; in the latter case, the installation did not pay actual shipping cost but an average charge, depending on mode of shipment.

[102]When considered from an overall Army perspective this statement may appear too strong. After all, quality of repair could affect ability to complete a mission. However, data are used to inform the decisionmaking processes that range from repair-or-buy to the amount of OMA funding allocated to each installation. These decisions also affect the ability to complete missions.

[103]While this was a one-time problem that has been fixed, we caution that EMIS is not a mature management information system and, as with any emerging system, can be expected to have flaws that must be corrected.

address some of the concerns that had arisen. The Army should anticipate that the maturation period for EMIS will continue for at least a few more years.

Many reports are available through EMIS, but more should be automatic. For example, if ISM is to achieve productivity gains, the various segments of the repair-and-transportation pipeline must be monitored. Automatic reporting of excess times in the system is essential.

The Army Must Exercise Caution on Cost-Avoidance Projections

The verdict on how much ISM will save the Army must be deferred until more-reliable data are available. The sample sizes of items actually repaired are small, so the variance in any estimate will be large. Furthermore, the lines repaired in ISM were specifically chosen to be the ones with the highest potential OMA cost avoidance; therefore, if linear extrapolations are used to estimate savings on other lines, savings are overestimated. Moreover, the use of retail credit rates rather than wholesale repair costs to evaluate cost avoidance may overstate savings for the Army as a whole. Some savings are likely to be realized over time, particularly if financial changes are implemented, but OMA budgets should not be reduced in the short term on the basis of overly optimistic projections.

Saving Money Requires Difficult Decisions

During the ISM PoP and ISM-X demonstration, the only way actual dollars could be saved was to reduce procurement, because the same number of people were employed at the DOLs and repair capacities were not reduced. In fact, based on discussions at the PP&Cs and our observations during site visits to COEs, there is evidence that DOLs are adding facilities and capabilities. Overhead functions have shrunk as a result of management initiatives, but not as a direct result of ISM. So the question really is, "How much money will be saved from reducing procurement?" No one can say, because the data are poor and because it is inappropriate to extrapolate demands from the nonrepresentative sample of lines used in the ISM PoP and ISM-X demonstration. However, it is possible to calculate accurately the effects of workforce reduction. True savings can come only when both procurement and infrastructure are trimmed.

The Army Should Integrate Velocity Management Methodology with ISM Expansion

Velocity Management is a program to improve the Army's logistics business, both in garrison and when deployed. Its bottom-line goal is to improve the effectiveness of the logistics processes in sustaining mission accomplishment. Ultimately, Velocity Management will result in reduced stocks and real dollar savings as the Army replaces logistics mass with precision and speed, providing a hedge against unforeseen interruptions in the logistics pipeline. Implementation of Velocity Management is going to ensure outstanding performance by finding and eliminating sources of delay and undependability in the Army's logistics processes. Under Velocity Management, the Army logistics community will continue to improve its support to the commanders in the field by measuring its performance closely. With the implementation of Velocity Management at the DOLs, processes should improve.

The Velocity Management methodology requires defining the process, measuring the process, and improving the process. ISM is also attempting to implement roughly the same three tasks, although, to date, not much emphasis has been placed on defining the process. ISM does measure—parts costs, number of days in repair, transportation times, man-hours, etc. ISM also is trying to improve the repair process by making it more efficient. Thus, there is potential for synergy between these two efforts, and it will work to the advantage of the Army in improving sustainment maintenance.

The Army Should Align Peacetime Practice With Contingency Doctrine and Practice

There are relatively straightforward ways to ensure that peacetime practices do not hamper the Army's ability to perform contingency operations. Among these are (1) conducting studies to capture demand data for components versus repair parts at the national item manager level; (2) establishing a close linkage between item managers, COE managers, and in-theater contingency maintenance managers; (3) ensuring that the AMC–Logistics Support Element (AMC-LSE) includes an NSMM member when deployed to a contingency; (4) incorporating training for theater army area command maintenance managers, especially those in Reserve Component organizations, into major Army and joint exercises; and (5) identifying organizational changes and allocating funds to procure necessary equipment and electronic information management system connectivity for both active and reserve component units that may deploy to a contingency (e.g., theater army area commands, corps support commands, and AMC-LSE).

WHAT ISM ACHIEVED

In spite of these significant problems, ISM achieved some noteworthy successes.

First, the Army has achieved a new visibility of maintenance capacity and capability. That alone may justify the immediate implementation of the evolving concept. Such visibility is absolutely indispensable for effective management of personnel and facilities.

Furthermore, some data suggest that there has been improvement. Overall, parts appear to get repaired more rapidly in the COEs under the ISM structure: 75 percent of the lines were repaired with fewer man-hours. Although participating installations compete seriously, a collegial environment is emerging among the installations as they attempt to institute more-effective repair programs. This collegiality has been enhanced by an inter-MACOM Corporate Board, which was established for the duration of the ISM-X demonstration. Those two groups are vital to developing new policies and procedures for improving the Army's maintenance programs.

Finally, repairing components across major Army commands was accomplished without excessive difficulty, something no one was sure could occur when the demonstration began.

Although ISM has accomplished some significant changes in logistics support, remaining problems must be resolved for it to achieve its full potential. Specifically requiring attention are the financial policies and incentives surrounding logistics operations, the management of the ISM program, support to contingency operations, and long-term decisions about the logistics infrastructure.

Appendix

A. ARMY FOUR-LEVEL MAINTENANCE STRUCTURE AND THE FLOW OF REPARABLES

DEFINITION OF FOUR-LEVEL MAINTENANCE

The Army maintenance system has four levels: operator/ organizational, direct support, general support, and depot. The maintenance at each level can be performed by soldiers, government civilians, or contractors. The decision on which level should perform a maintenance action is based on a variety of criteria, but primarily the amount of time that maintenance action takes, the criticality of the repair to a unit's operational mission, and the requirement for any special tools, facilities, or test equipment.

The operator/organizational level takes care of the preventive-maintenance tasks, as well as the simpler replacement of components—for example, oil changes, tire replacement, spark plug replacement, and other simple on-vehicle tasks.

The direct-support level involves more-complicated tasks that require either more time or some special test equipment. Such tasks involve the removal and replacement of an engine or the replacement of certain components on an engine or transmission, which may require some engine disassembly. At this level, support involves repair by replacing components on an item and returning that item to the original user. It may also involve repairing a component or an item and returning it to the direct-support unit's supply support activity.

The general-support level includes tasks at the direct-support level performed as a backup to the direct-support organization, as well as tasks that are more complicated. For example, general-support maintenance on a vehicle may involve a complete transmission rebuild, or rebuilding components that are part of the transmission. It may also involve repairing high-density components that are better repaired on a production line, rather than as the individual failures occur.[104] In most cases, the general-support level returns repaired items to the general-support unit's supply support activity, not to the user.

Depot-level maintenance involves production-line repair of components and end-item overhaul. The workload could come from a variety of sources: overflow work from the general-support supply system in the field; components taken from end items being rebuilt; or turn-ins resulting from National Inventory Control Point (NICP) instructions about the disposition of unserviceable reparables. Depot operations support the national wholesale

---

[104]Theoretically, all ISM items are high-density items.

supply system. The NICP item manager decides whether to procure new items or to repair unserviceable reparables. This decision is made based on several variables, including cost to repair versus cost to procure, available sources of supply, and timeliness of procurement versus repair. If the decision is to repair, then the NICP item manager selects the organization that is believed to be most appropriate for meeting supply-system requirements. This repair could be done by a government depot, by a contractor, or by one of the ISM Center of Excellence (COE) sites in a bidding process as described in the main body of this report.

FLOW OF SERVICEABLE REPARABLES FROM POINT OF FAILURE TO POINT OF REPAIR

ISM deals primarily with the repair of unserviceable reparable components. The flow of unserviceable reparable exchange components in peacetime typically follows the path shown in Figure A.1. The first step in the process is to remove the component at the point of failure, which can be done at any of the four maintenance levels. For simplicity, we begin with a component that is removed at the operator/ organizational level. However, an unserviceable component could have been turned in to the supply system by any of the maintenance organizations elsewhere, such as DS, GS, or depot.

Figure A.1 illustrates that the flow of unserviceable reparables under ISM is a supply-system activity. The supply organization determines whether to obtain replacement components either by local repair programs or by requisitioning from higher-level sources of supply. This decision is made at each level. Once a decision to repair a line is made, instructions are sent to the supported organizations to turn in unserviceable items to the supply support activity. If DS and GS supply organizations have an ongoing repair program for a component, then that component is put on a reparable exchange (RX) list. Reparable exchange means that when an item fails and a replacement is ordered, the replacement is supposed to be provided on a one-for-one exchange by the supply-support activity. That activity's maintenance program is supposed to be responsive enough to have replacement components available for issue upon demand. When the maintenance program is not robust enough, either the customer has to wait or the supply-support activity orders an item from its next-higher source of supply.

At each supply organization, an inspection is made to determine whether a particular unserviceable reparable component should be repaired or discarded. For example, even though the general set of items is supposed to be repaired, any particular item turned in to the supply system may not be in good enough condition to repair because the damage is so

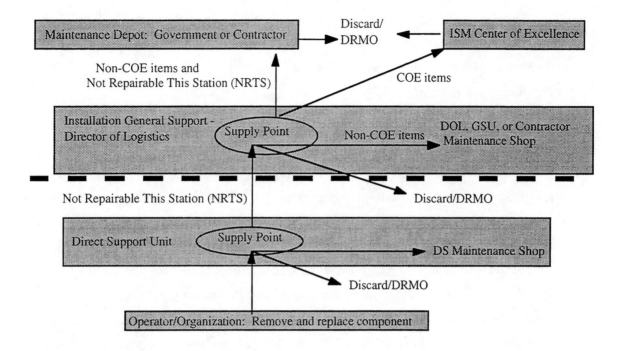

Figure A.1—Typical Peacetime Flow of Unserviceable Reparable Exchange Components

extensive as to be uneconomically reparable. In that case, the item is sent to the Defense Reuse and Marketing Office (DRMO) at the servicing installation or discarded, depending on the instructions from the NICP.

In some cases, the item may not be reparable at a particular level, and so it is declared "Not Reparable This Station" (NRTS) and is sent to the next-higher level.

If there is an ISM repair program for a particular item, the item is sent to the ISM Center of Excellence (COE) for repair. The COE may or may not be at the installation servicing the supply organization to which the reparable had been turned in. If the COE is at the same installation, the item is repaired and then put into installation stock. If the COE is at a different installation, the item is sent to that installation, repaired, and then returned to the original installation's stock.

Figure A.1 does not depict the flow of "ISM national work," i.e., items that are sent to a depot for repair and require GS-level repair tasks. Such work may be done at a government depot, a contractor, or an ISM COE. If the ISM COE does the repair, there is a separate process for bidding through the National Sustainment Maintenance Manager (NSMM), but the repairs are done at a COE just as are those arriving from non-COE organizations. The extent of repair may be different, as agreed between the NICP and the NSMM.

The heavy dashed line in Figures A.1 and A.2 separates the maintenance levels into two groups. Below the line is readiness maintenance, and above the line is sustainment maintenance. Because COE repairs occur above the heavy line, they are termed "sustainment maintenance."

FLOW OF SERVICEABLE REPARABLES FROM POINT OF REPAIR TO SUPPLY SYSTEM

All items repaired by a COE or by local repair programs are normally turned in to the supply system for reissue. For local repair programs, the condition of the item after repair warrants that the item be issued only within the servicing installation, for several reasons: The repair is done on a limited basis (inspect and repair only as necessary), the repaired item is not packed sufficiently for long-term storage or shipment to a variety of geographic/environmental conditions, and its use in different geographic/environmental areas may be restricted. Items that are repaired under the national workload program are repaired and packed to different standards, which are specified by the wholesale supply system for meeting general-issue standards.

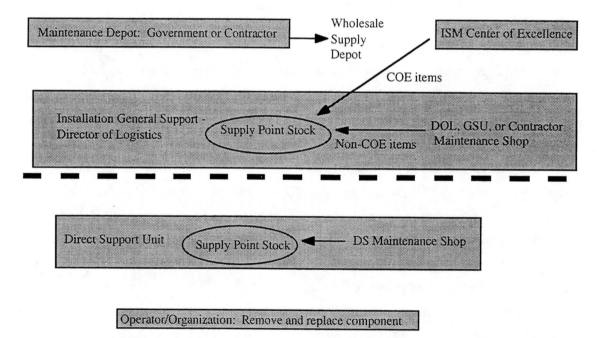

Figure A.2—Typical Peacetime Flow of Repaired Reparable Exchange Components

B. ISM-X LINES

COE	NIIN	Nomenclature	AMDF Price ($)
Bliss			
	00-304-3493	STARTER ENGINE M939	393
	00-971-5016	XMSN SP645 M800 SERIES	4741
	01-120-2169	TRANSMISSION M113A2	5463
	01-146-5483	TRANSFER CUCV	720
	01-171-4774	ELECTRONIC ASSY M1	10321
	01-179-7523	OPERATOR'S PANEL M1	4631
	01-214-0988	MULTIPLEXER MULTI	11731
	01-259-0807	PROCESS AY APR39	11760
	01-266-5964	RADIO SET RADIO	910
	01-270-2982	HUB ROTOR UH1	52035
	01-301-8212	STABILIZER HORIZONTAL UH60	54019
Carson			
	00-001-4076	CONTROL RADAR	3036
	00-058-1111	MODULE AVIONICS	523
	00-086-7792	XMSN MDL CD8506A AVLB/CEV	75347
	00-138-8687	REGULATOR MULTI	747.66
	00-140-7531	XMISSION M88	91718
	00-178-0268	ENGINE 5 TON	22958
	00-530-8742	ENGINE DIESEL GEN SET	5642
	00-740-9589	POWER TAKEOFF M54	858
	00-872-5971	ENGINE 2A0 GEN 10 HP	3070
	00-896-9020	FINAL DRIVE L M88	12334
	00-896-9021	FINAL DRIVE R M88	19268
	01-016-3572	SERVO AH1	1693
	01-021-3288	RECEIVER RADAR	6713
	01-046-5861	ENGINE DIESEL GEN 5KW	4967
	01-047-8613	TRANSFER XMSN M548	2935
	01-065-3716	REGULATOR UH1	669.86
	01-065-3979	PUMP FUEL M915	1583.66
	01-073-0056	TRANSFER M915	822
	01-075-2735	CONTROL ASSY M915/16/2	3729
	01-082-8143	TRANSMISSION M915/16/20	20753
	01-087-0156	TRANSFER T RPI (OFF) M113A2	1772
	01-092-7956	CIRCUIT CARD RADIO	178
	01-104-1446	FUEL PUMP M809	1482.9
	01-111-2262	ENGINE DIESEL M939	22944
	01-129-6508	HUB ASSY OH58A/C	14740
	01-141-1257	BLOWER HEMTT	2174.49
	01-141-9372	PUMP FUEL M915A1	1456
	01-142-2745	ENGINE DIESEL M915A1	26421
	01-150-5002	ENGINE DIESEL CUCV	7010
	01-210-0429	CLUTCH FAN HEMMT	870.83
	01-218-2206	TRANSMISSION TRACTOR	18395
	01-230-3922	CYLINDER HEMTT WRK	922.03
	01-264-6542	GENERATOR HMMWV	1421
	01-268-8757	INJ PUMP M939A2	1303

COE	NIIN	Nomenclature	AMDF Price ($)
Hood			
	00-043-1987	AMP OSCILL AVIONICS	5490
	00-089-8287	TRANSFER CASE M35	1867
	00-116-8241	PUMP FUEL M35	2951
	00-134-4803	ENGINE UH1	152712
	00-167-9757	GYRO NAV AVIONICS	5738
	00-181-1955	INDICATOR UH1	1565
	00-457-0571	RECEIVER/TRANSMITTER TACTICAL COMMO	349
	00-501-7001	ENGINE DIESEL GEN SET	7925
	00-550-6628	GYRO AVIONICS	8174.76
	00-621-1860	ENGINE AIRCRAFT AH1	149347
	00-841-0302	INDICATOR UH1	2984
	00-853-5915	OSCILLATOR TACTICAL COMMO	203
	00-908-6320	PUMP METER M54	1739
	01-013-1339	ENGINE AIRCRAFT T63	146481
	01-030-4890	ENGINE AIRCRAFT CH47	632754
	01-046-5862	ENGINE DIESEL GEN 10K	7112
	01-070-1003	ENGINE UH60	591467
	01-070-3660	ATTITUDE INDICATOR OH58	21752
	01-074-3488	OIL PUMP ASSY M1/A1	6075
	01-076-6741	ELEVATING MECH M1/A1	4340
	01-091-1659	AXLE ASSY M939	6168
	01-106-1903	BLADE UH60	84796
	01-107-9930	GENERATOR M2/3	2873
	01-109-7969	ENGINE MK4FL	7810
	01-109-8824	INDICATOR TACH OH58A/C	2475
	01-113-8188	BLADE ROTARY UH60	48825
	01-114-2211	ENGINE AIRCRAFT AH64	549461
	01-114-8608	PUMP FUEL COMMON	801.33
	01-118-2885	RADIATOR M2/3 BFV	2467
	01-128-9544	ENGINE M911	17091
	01-134-6531	POWER PACK M2/3	1505
	01-135-6251	TRANSMITTER ARC164	2974.47
	01-137-8137	BLADE OH58A/C	13759
	01-160-3475	INDICATOR AH64	2960
	01-160-8008	QUADRANT ASSY COMMON USE	4673.61
	01-160-8044	INDICATOR AH64	3551
	01-161-1200	INDICATOR TORQUE H AH64	5537
	01-161-2136	TRANSMISSION HMMWV	2224
	01-161-3919	ATTITUDE HEAD (HARS) AH64	80766
	01-162-6978	PANEL FAULT AH64 F	6384.43
	01-163-7133	CIRCUIT CARD AH64	709
	01-166-2051	ENGINE CEV/AVLB	128802
	01-167-4282	FINAL DR M1/M1IP	7099
	01-171-2381	CIRCUIT CARD M1/M2/M3	853
	01-172-2886	SYMBOL GENERATOR AH64	18574
	01-172-3055	CIRCUIT CARD M1/M2/M3	1235
	01-178-7479	ANGLE DRIVE M1	5753
	01-183-5499	RADIO MAG AH64	4416
	01-199-2049	TV SENSOR TADS AH64	43430

COE	NIIN	Nomenclature	AMDF Price ($)
	01-199-2355	PUMP FUEL HMMWV	585
	01-204-4470	GENERATOR M2/3	3529
	01-210-8795	XMISSION M1/M1IP	231556
	01-211-0130	MRTU TYPE 1 MULTI	28894
	01-211-6346	DISPLAY M139	8274
	01-213-1623	HOIST ASSY HEMTT	5948.9
	01-214-0201	BLADE ASSY.TAIL OH58A/C	2751
	01-214-2640	NOZZLE ASSY M1/M1IP	928
	01-218-2208	POWER TAKE OFF M2/M3 BFV	7321
	01-220-9395	STATION DIR STATION	3189
	01-224-4838	PROCESSOR AH64	3417
	01-226-7509	CONTROL DIR AH64	14449
	01-232-4442	SIGHT UN M139	28277
	01-232-6568	DAY SENSOR AVIONICS	146632
	01-232-6702	ELEC UNIT L TADSAH64	30453
	01-237-5490	PANEL ASSY MULTI	4578
	01-237-8098	DISPLAY UNIT AVIONICS	22906
	01-239-1616	CONVERTER SIGNAL AH64	14526
	01-241-7043	SIGNAL PROCESSOR AH64	3508
	01-245-9091	RECEIVER TRANSMITTER AVIONICS ALTMTR	19797
	01-245-9093	INDICATOR HEIGHT ALT IND	6629
	01-245-9094	RECEIVER ALTIND	11315
	01-254-7793	GEAR BOX AH64	37582
	01-257-3879	ENGINE DIESEL W/CONT HEMTT	31039
	01-257-3880	TRANSMISSION W/CONT HEMTT	13199
	01-258-7001	CONTROL AH64	6439
	01-258-7074	ELECT ASSY TADS AH64	54546
	01-259-0154	ELECTRONIC UNIT AAH MICOM	11686
	01-260-0212	ENGINE W/C M109A5	21867
	01-261-4444	SWITCH M139	29112
	01-263-1815	MRTU TYPE 3 MULTI	30296
	01-265-6947	COMPUTER FC AH MISSILE SYS	114567
	01-271-0297	PUMP HYDRA M2/M3 BFV	305
	01-275-8092	CIRCUIT CARD M1/M2/M3	2067
	01-282-2846	CONTROL ECCM RT1523	1154
	01-284-5722	FAN VANEAXIAL M109	1211
	01-290-1290	ENGINE 600HP W/CONT M2/3A2	48803
	01-291-9334	HELMET DISPLAY UNIT AH64/M139	11761
	01-293-9706	THERMAL RECEIVER M1	94218
	01-295-7458	ENGINE DIESEL M113A2	10336
	01-299-6100	PWR SUPP TP TADSAH64	53034
	01-301-3226	ELEC UNIT P TADSAH64	50862
	01-305-6955	PUMP HYD AH64	4331
	01-307-9447	TURRET SENSOR SIGHT AVIONICS	192854
	01-308-3019	TURRET TADSAH64	220022
	01-312-2387	BLADE AH64	18467
	01-314-7940	ENGINE W/C HMMWV	6451
	01-316-6617	ENGINE W/C M551	19704
	01-317-9799	ELECTRONIC CONTROL M1A1	18898
	01-332-0702	BLADE ROTARY WING AH64	99797

COE	NIIN	Nomenclature	AMDF Price ($)
	01-380-0280	SIGHT UNIT M1A1	63482
	01-381-1842	SIGHT UNIT M1A1	61660
	01-384-5304	PUMP FUEL M2/M3 BFV	1589
	01-392-3735	ELECTRONIC COMP TADS/LRU	125309
Kansas National Guard			
	00-641-6405	WINCH DRUM M809	2029
	00-753-8687	WINCH DRUM M35A2	2029
	01-225-1031	PUMP FUEL COMMON USE	2214.1
Riley			
	00-106-7754	TELESCOPE PANORAMIC M109	8906
	00-124-5387	ENGINE M88	119965
	00-712-1280	ANGLE DRIVE COOLING M113/FOV	500
	00-830-6660	GENERATOR AVLB/CEV	2049
	00-884-2492	MODULE PUSH BUTTON TAC RADIOS	752
	00-972-2638	CYLINDER LIFT TRUCK 5T	591
	00-972-2639	CYL BOOM M816	1239
	00-974-9670	BOOM HOIST WINCH M939	2810
	01-067-3842	IMPELLER FAN AXIAL M1	1457
	01-069-0483	DRIVE UNIT M1/M1IP	805
	01-073-0124	PMP FUEL M88	4042
	01-076-6740	GRIP ASSY M1	2035
	01-076-6865	GRIP GUN M1A1	2309
	01-084-3447	ENGINE DIESEL M915/16/20	31794
	01-089-4896	FINAL DRIVE M2/3	3084
	01-092-8067	PWR SUPPLY FM RADIOS	554
	01-105-6445	ENGINE 500HP RPI (OFF) M2/3 A1	40634
	01-117-3010	TRANSMISSION W CONT M939	12786
	01-139-3722	STARTER M60	895
	01-144-1528	TRANSFER W/CONT M939	5122
	01-146-5482	TRANSMISSION CUCV	2145
	01-151-2684	TURBOCHARGER M109	840
	01-154-9993	FINAL DRIVE M9 ACE	5272
	01-163-4999	TRANSFER XMSN HMMWV	1673
	01-214-8820	ENGINE M35	14596
	01-215-6721	PUMP FUEL GENERAL USE	683.45
	01-217-2331	SWITCHING REGULATOR M1/2/3	2298
	01-217-2334	CONVERTER ASSY M1A1/M2	3999
	01-217-8309	PUMP FUEL 500HP M2/3	1656
	01-238-8186	MANUAL DRIVE M1/M1IP	3026
	01-240-8789	FINAL DRIVE M2/3	5309
	01-246-1873	ELECTRONIC UNIT M1/M1A1	12048
	01-264-2040	LASER RANGEFINDER M1	29199
	01-275-7477	GENERATOR M1/M1IP	7503
	01-276-5946	ENGINE GAS MSE 6HP	2481
	01-316-9270	TURRET NET BOX TADS/LRU	20153
	01-339-4716	TRANSMISSION W/CONT M9 ACE	39959
	01-348-1332	TRANSFER HEMTT	6393
	01-351-6549	ENGINE W/CONT M9 ACE	44638
	01-374-7539	ENGINE W/CONT M939A2	17628

COE	NIIN	Nomenclature	AMDF Price ($)
Sill			
	00-693-0617	STEERING GEAR M35	708
	00-714-6135	DIFFERENTIAL STEERING M113A2	5633
	00-894-9533	TRANSMISSION 411A2A M109	66866
	00-894-9535	TRANSFER M109	17924
	00-964-9203	FINAL DR W/C M109	4424
	01-051-1679	TRANSMISSION HET	16861
	01-106-8608	BOOM CONTROL MLRS	730
	01-117-7232	PANEL ASSY PADS	15084
	01-118-2885	RADIATOR M2/3 BFV	2467
	01-120-4131	COMPUTER PADS	42585
	01-139-6339	POWER SUPPLY PADS	52306
	01-214-3315	TRANSDUCER MLRS	3250
	01-240-0639	IED MLRS	10144
	01-274-6449	XMISSION M2/M3 BFV (600HP)	97697
	01-276-5947	ENGINE MIL STD ENGINE	4164
	01-288-0497	ALTERNATOR M109	1910
	01-305-7685	CONTROL PANEL MLRS	24902
	01-336-9616	ELECTRONIC UNIT MLRS	125320
	01-338-1703	PAYLOAD INTERFACE MLRS	60740
	01-338-2703	TRANSMISSION M2/M3 BFV(600HP)ELEC	189223
	01-346-7913	POWER DISTRIBUTION MLRS	5554
	01-374-5129	FIRE CONTROL UNIT MLRS	82226
	01-389-7158	FIRE CONTROL BOX MLRS	34793
Texas National Guard			
	00-430-3480	ENGINE GENERATOR	9786
	00-706-1137	POWER TAKEOFF M35	666
	01-058-1161	FINAL DRIVE AVLB/ CEV	5801
	01-061-5766	FINAL DRIVE M113	1336
	01-096-5151	SIGHT ASSY TOW M2/3	22829
	01-107-9902	FAN VANE M113	1527
	01-260-0212	ENGINE W/CONT M109	21867
	01-273-5946	CONTROL BOX M2/3	33042
	01-341-4647	FUEL PUMP INTANK M1A1 TANK	413

COE   Center of Excellence

NIIN   National Inventory Identification Number

AMDF   Army Master Data File

C. DESCRIPTION OF DATA USED TO ASSESS COE REPAIR PERFORMANCE

For the purposes of our analysis, we defined the ISM-X demonstration period to be from July 1, 1995, to January 2, 1996. The start date was picked as a time when all ISM-X Centers of Excellence (COEs) had received their EMIS (Executive Management Information System) equipment and were online and fully participating. We consider ISM-X repairs to be those started in this time window at the appropriate COE on ISM-X lines, i.e., those lines continuing in repair at the COEs from previous Planning, Production, and Control (PP&C) conferences, as well as those added at PP&C 5 and 6 (we explicitly excluded repair work on ISM-X lines that was done for the Army Materiel Command's (AMC's) Major Subordinate Commands). Consistent with ISM-X practice, we aggregated items to their prime National Inventory Identification Numbers (NIINs). Although there are differences in parts used to repair different substitutables, ISM-X COEs gave one bid for all substitute NIINs. To get a baseline for repair-process measures, we used repairs initiated on ISM-X lines in the nine months prior to the start of the ISM-X demonstration, roughly October 1, 1994, through June 30, 1995.

We acquired the data from one of the primary data-collection efforts in ISM-X: the databases maintained in the EMIS run by the Regional Sustainment Maintenance Manager (RSMM) at Fort Hood for day-to-day management of ISM—monitoring the performance of COEs, identifying problems, checking on problems with ISM-X lines at supported units, etc. EMIS collects maintenance data from Army "legacy" data systems (such as MIMS and SAMS)[105] and the EMISs at each individual COE. It also includes some special databases such as transportation information that is collected directly.

We used two of the data files maintained by EMIS in our analysis. The first was the EVAC file, which maintains transportation data (and some work order information) on ISM-X lines that are transported between installations. The EVAC file is cumulative, beginning with the start of the ISM Proof of Principle. We also used the WORKORDERS file, which contains extracts from individual jobs from each installation's maintenance data systems. The current WORKORDERS file contains jobs that were open at the start of the current fiscal year (October 1995) or were opened afterwards. We acquired a copy of each of

---

[105]MIMS (Maintenance Information Management System) is used by the installation DOL. SAMS (Standard Army Maintenance System) is used by general-support (GS) and direct-support (DS) maintenance units.

these files each Monday morning from the RSMM at Fort Hood.[106] The RSMM staff also provided us with a copy of the WORKORDER archive file for 1995, which contained work orders closed as of the end of the fiscal year.[107] The analyses in Section 3 are based on the EVAC and WORKORDER files as of January 2, 1996 plus jobs from the archive file that were begun in FY95.

We kept both open and closed jobs in our data set. As noted in Section 3, we used only closed jobs or a mixture of open and closed jobs, depending on our analysis goals. In general, for our analyses of repair processes, we used those jobs that had a repair quantity of one or more (i.e., we omitted jobs for which items received were condemned or declared not reparable[108]).

Our analysis data set had 12,249 jobs for ISM-X lines (13,000 items returned). Of these, 7,304 jobs (7,782 items) were from the baseline period before ISM-X, and 3,680 jobs (3,843 items) were ISM-X items repaired during ISM-X at their respective COEs.[109]

---

[106]We copied the file from Fort Hood electronically over the Internet to our workstation at RAND.

[107]We noted above that our comparison group of jobs contained only those starting on or after the start of the 1995 fiscal year, even though the archive file contains jobs that were open when the fiscal year began. We eliminated those latter jobs to avoid the potential for length bias.

[108]These items are declared "NRTS" (Not Reparable This Station), which means that while the item may be reparable, the present maintenance organization does not have the authorization to repair it. The item is then transported to an organization that does have the authorization to make the repair.

[109]There were 4,945 jobs (5,218 items) involving ISM-X COE lines during the ISM-X demonstration period. The non-COE repairs were the continuation of work in progress at those facilities when ISM-X began or simple repairs for which it was not deemed necessary to send the item to its COE. All of these jobs are collected in EMIS.

D. HOW TO INTERPRET A BOXPLOT

Figure D.1 represents a generic boxplot like those found in Section 3. The vertical axis represents the ratio of the median before the start of the demonstration (old median) and the median during the demonstration (new median). A ratio of 1 indicates that the two medians are the same, i.e., median performance has not changed. A ratio greater than 1 indicates that performance degraded during the demonstration; a ratio less than 1 indicates improved performance. The y-axis is logarithmic, not linear.

- The ratio is bounded below by zero (the median days, man-hours, and parts cost cannot be less than zero), but can be any value above zero. With a linear scale, when the median value improved during ISM-X (a ratio less than 1), the points corresponding to improvements are relegated to the region between zero and one. The logarithmic scale gives equal weight to both improvements and degraded performance (values above 1).

- Second, with a logarithmic scale, a ratio of 5 to 1 (an increase in the measure to 5 times its pre-ISM-X value) is the same distance above 1 as a ratio of 1 to 5 (a decrease in the measure to one-fifth its pre-ISM-X value) is below 1.

- Finally, the logarithmic scale allows us to present the full range of data on a single graph, even with ratios as high as 40 and as small as 0.1. The values on the y-axis give the actual ratio values.

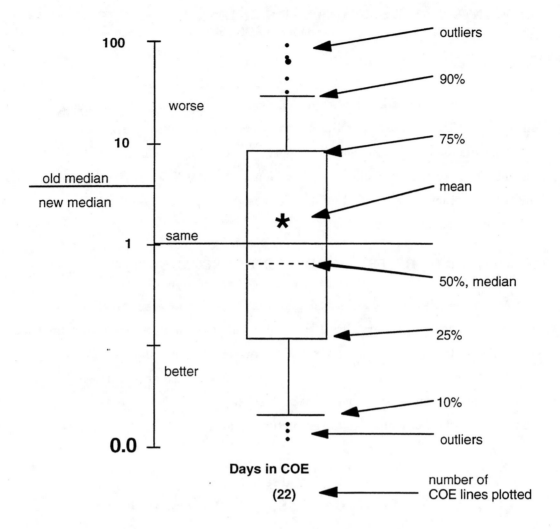

Figure D.1—Illustrated Boxplot

E. INTEGRATED SUSTAINMENT MAINTENANCE AND THE RESERVE COMPONENTS

This appendix provides additional background on Integrated Sustainment Maintenance (ISM) with regard to the Reserve Components. It is divided into three parts: (1) background information on general-support (GS) maintenance in the Reserve Components, (2) reserve component participation in the ISM-X demonstration and possible future participation in Armywide ISM, and (3) possible implications of ISM for the Reserve Components in contingencies.

BACKGROUND ON RESERVE COMPONENT GENERAL-SUPPORT MAINTENANCE

Approximately 75–80 percent of deployable GS personnel are in the Reserve Components. The Army Reserve has 12 GS units and between 60 and 310 personnel slots (see Table E.1); the National Guard has 36 GS maintenance companies.

Army Reserve GS units do not necessarily collocate with active units, so they generally do not have equipment or opportunities to perform GS repair. The Army Reserves do not have full-time people or embedded infrastructure to do maintenance; they must, therefore,

Table E.1

Army Reserve General-Support Maintenance Units

Unit	Personnel Authorizations	Location
238th	90	San Antonio, TX
417	88	Faribault, MN
344	122	Bogalusa, LA
1007	310	Hargerstown, PA
928	81	Macon, GA
279	90	Dallas, TX
310	123	Jackson, MI
682	83	Canton, OH
304	122	Bartlesville, OK
318	79	State College, PA
424	102	Rolla, MO
489	63	San Bernardino, CA

SOURCE: Office of the Chief, Army Reserve (OCAR), Logistics Division, Maintenance and Materiel Branch.

depend on others to do most of their maintenance. The Army Reserve does only between 3 and 11 percent of its own maintenance. In the past, it relied largely on the active Army installation Directorates of Logistics (DOLs). In short, GS repair does not constitute much work for the AR.

The National Guard, on the other hand, performs a great deal of GS-level work[110]—some at Combined Support Maintenance Shops (CSMSs), but most at Mobilization and Training Equipment Sites (MATES). The National Guard has 24 MATES around the country (see Table E.2). MATES are usually authorized 120 personnel but typically fill only about 78 positions.

Table E.2

National Guard Mobilization and Training Equipment Sites

Name	State
MATES for Alaska	AL
Camp Roberts MATES	CA
Fort Irwin MATES	CA
Fort Carson MATES	CO
Camp Blanding MATES	FL
Fort Stewart MATES	GA
Hawaii MATES	HI
Gowen Field MATES	ID
Fort Knox MATES	KY
Fort Riley MATES	KS
Camp Beauregard	LA
Fort Polk MATES	LA
Camp Grayling MATES	MI
Camp Ripley MATES	MN
Camp Shelby MATES	MS
Fort Drum MATES	NY
Dona Ana MATES	NM
Fort Bragg MATES	NC
Fort Sill MATES	OK
Puerto Rico MATES	PR
Fort Hood MATES	TX
Fort Pickett MATES	VA
Yakima MATES	WA
Fort McCoy MATES	WI

SOURCE: National Guard Bureau (NGB), May 1995.

---

[110]Conversations with personnel assigned to the Logistics Division of the OCAR, September 1995, and the National Guard Bureau, January 1996.

There are a few light brigade MATES (e.g., Florida), but most MATES house heavy equipment. MATES are manned by military technicians (or state technicians) who are full-time civilian employees paid for by federal funds. However, they must also belong to the Reserves or National Guard and, therefore, might deploy with some other unit in a contingency. Because National Guard equipment is centralized at the MATES, these reservists have ample opportunities to perform GS-level work. This volume of work also enables them to take in GS-level repair from outside the National Guard. In addition, their parent unit might be a direct-support (DS) unit, which enables them to work on different equipment at GS and DS levels, as they might in wartime.

THE RESERVE COMPONENT, ISM-X, AND ISM

The Army Reserve participated in ISM-X planning and Corporate Board meetings, but it did not participate in ISM production, in large part because USAR does very little GS-level work and because it was in the midst of a major reorganization. However, questions still remain about the level and extent of its participation as ISM is implemented Armywide.[111]

To enable USAR participation, the Regional Sustainment Maintenance Manager (RSMM) could divert certain items to USAR GS units for their annual training drills. Or USAR might participate in ISM strictly as a customer, along the lines of National Training Center (NTC) participation, rather than as part of a production line.

A fertile subject for further exploration of possible USAR contributions to ISM is the operation of the Kaiserslautern Industrial Center by the USAR in Germany. Initial inquiries suggest that it is an example of a USAR facility providing continuous GS-level maintenance support to the active Army.

The National Guard, by contrast, participated in ISM-X planning and production and intends to increase its participation as ISM expands. The Texas and Kansas National Guards competed for and won Center of Excellence (COE) lines, but the Kansas National Guard performed very few repairs under ISM-X. Attention should be directed toward future ISM participation by the NG.

The MATES, in particular, are well suited to participation in ISM, for several reasons. First, MATES have enough GS capacity to enable them to compete for certain COE lines. Second, GS repair capacity at MATES can expand as needed, because the NG does not face the same personnel restrictions as the active Army, such as civilian personnel limits at installations (see Section 4). Third, the military technicians who run the MATES GS repair

---

[111]As of the writing of this report, Logistics Support Activity (LOGSA) has informed us that two USAR posts—Forts Dix and McCoy—are planning to participate in ISM repair.

are also in the reserves, which serves ISM's objective of expanding training opportunities for the Reserve Components. Greater utilization of MATES in ISM will expand the training benefits to the reserves.

Outstanding issues remain about how the National Guard GS management structure will fit into the Armywide ISM management structure and how workload is allocated to MATES with COE lines. Nevertheless, the National Guard is likely to continue as an active participant in ISM.

THE RESERVE COMPONENT, ISM, AND CONTINGENCIES

As stated in Section 1, one objective of the ISM concept was to improve sustainment maintenance in contingencies by improving coordination and visibility of maintenance assets and activities. Previously, a theater commander had to coordinate sustainment-maintenance requirements separately with Army staff, U.S. Army Forces Command (FORSCOM), Army Materiel Command (AMC), and other Major Army Commands (MACOMs), as well as the National Guard Bureau (NGB) and OCAR. At this point, it is not known how future ISM will integrate NGB and OCAR management of sustainment maintenance, although, as discussed above, it has clearly improved the visibility and coordination of sustainment maintenance in the active Army.

Moreover, the organization and management of GS units in a contingency are also in flux as GS units shift from the corps command to the theater army command. The implications of these changes for the Reserve Component GS units and resources require further study.

Additional uncertainty surrounds how the GS shops in the continental United States (CONUS) will operate if reserve personnel from, for example, the MATES are deployed. The Texas National Guard (TXNG) MATES envisions a relatively smooth continuation of GS maintenance, even if most of the full-time military technicians deploy, because it has a ready pool of additional skilled labor.[112] According to the NGB, however, other MATES might shut down operations entirely, possibly shifting work to another state. These possibilities have implications for the Army, depending on the types of equipment being repaired at the various MATES.

The USAR, according to OCAR, would fill in GS slots as needed and might even operate at higher productivity levels because units would have fewer "distractions." Of

---

[112]In Operation Desert Storm (ODS), if the 49th Armored Division had deployed, 40 percent of the MATES workforce would have deployed. The MATES would have hired temporary employees to fill in.

course, all of these concerns are a function of the size and duration of the contingency. In anything less than a major regional contingency, GS operations in CONUS might change very little.

In sum, a great deal of uncertainty exists with regard to the participation of the Reserve Components in general-support maintenance in a contingency. This uncertainty is exacerbated by the introduction of a new concept for sustainment maintenance. As the Army proceeds with the implementation of ISM, further study should determine what, if any, changes are required to Reserve Component GS personnel, organization, and resources.

REFERENCES

Bolten, J., J. Halliday, E. Keating, and J. Sollinger, "FORSCOM Installation Cost Project," unpublished RAND research.

Bolten, J., J. Halliday, and E. Keating, *Understanding and Reducing the Costs of FORSCOM Installations*, Santa Monica, CA: RAND, MR-730-A, 1996.

Brauner, Marygail, "Evaluation Plan for the Army's ISM-X Demonstration," unpublished RAND research.

Brauner, Marygail, and John R. Bondanella, *ISM-X Evaluation: Briefing to ISM Corporate Board*," unpublished RAND research.

Brauner, Marygail, and J. R. Gebman, *Is Consolidation Being Overemphasized for Military Logistics?* Santa Monica, CA: RAND, IP-103, 1993.

Brauner, Marygail, James S. Hodges, and Daniel A. Relles, *An Approach to Understanding the Value of Parts*, Santa Monica, CA: RAND, MR-313-A/USN, 1994.

Brauner, Marygail, John Bondanella, Joe Bolten, Lionel Galway, and Ellen Pint, "ISM-X Preliminary Findings: Briefing to ISM Corporate Board," unpublished RAND research.

Brauner, Marygail, John Bondanella, Joe Bolten, Lionel Galway, Ellen Pint, and Rick Eden, "ISM-X Evaluation and Policy Implications for ISM Implementation," unpublished RAND research.

Cohen, Irving, *Coupling Logistics to Operations to Meet Uncertainty and the Threat (CLOUT): An Overview,* Santa Monica, CA: RAND, R-3979, 1991.

Computer Systems Development Corporation, *Integrated Sustainment Maintenance Proof of Principle Final Evaluation Report*, October 1994.

Crawford, Gordon, *Variability in the Demands for Aircraft Spare Parts*, Santa Monica, CA: RAND, R-3318-AF, 1988

Donohue, George, and Marygail Brauner, *Revamping the Infrastructure That Supports Military Systems*, Santa Monica, CA: RAND, IP-140, 1993.

Galway, Lionel A., "ISM-X Evaluation: Individual Item Repair Performance," unpublished RAND research.

——, *Management Adaptations in Jet Engine Repair at a Naval Aviation Depot in Support of Operation Desert Shield/Storm,* Santa Monica, CA: RAND, N-3436-A/USN, 1992.

Galway, Lionel A., and Christopher H. Hanks, "Data Quality Problems in Army Logistics: Classification, Examples, and Solutions," unpublished RAND research.

Unpublished RAND research from the ISM-X evaluation is available to Army clients.

Gebman, J. R., D. W. McIver, and H. L. Shulman, *A New View of Weapon System Reliability and Maintainability,* Santa Monica, CA: RAND, R-3604/1-AF, January 1989.

Girardini, Kenneth, et al., "Measuring Order and Ship Time for Requisitions Filled by Wholesale Supply," unpublished RAND research.

Girardini, Kenneth, and Louis Miller, "Metrics for the Army's Stockage Determination Processes," unpublished RAND research.

Gitlow, Howard, Shelly Gitlow, Alan Oppenheim, and Rosa Oppenheim, *Tools and Methods for the Improvement of Quality*, Homewood, IL: Richard C. Irwin, 1989.

Headquarters, Department of the Army, *Logistics Support Element Tactics, Techniques, and Procedures*, Initial Draft, Field Manual 63-11, undated, circa 1995/1996.

———, *Corps Support Command*, Field Manual 63-3, September 1993.

———, *Combat Service Support*, Field Manual 100-10, October 1995.

———, *Army Operational Support*, Field Manual 100-16, May 1995.

Hodges, James, *Onward Through the Fog: Uncertainty and Management Adaptation in Systems Analysis and Design,* Santa Monica, CA: RAND, R-3760-AF/A/OSD, 1990.

Miller, Rupert G., *Survival Analysis*, NY: Wiley, 1981.

Ogden, George C., and David M. Robinson II, *Development of Full Cost Solutions for the Integrated Sustainment Maintenance Proof of Principle,* Alexandria, VA: MPRI, September 10, 1993.

Pint, Ellen M., "Economic Incentives in the Use of Army Stock Funding," unpublished RAND research.

———, "Unfettered Buyers and Constrained Sellers, or Will ISM Save the Army Money?" unpublished RAND research.

Robbins, Marc, "The Need to Measure Repair Cycle Time," unpublished RAND research.

———, *Developing Robust Support Structures for High Technology Support Systems: The AH-64 Apache Helicopter*, Santa Monica, CA: RAND, R-3768-A, 1991.

——— "Improving the Army's Repair Process: Baseline Repair Cycle Time Measures," unpublished RAND research.

Rogerson, William P., *On the Use of Transfer Prices Within DoD: The Case of Repair and Maintenance of Depot-Level Reparables by the Air Force*, McLean, VA: Logistics Management Institute, PA303RD1, March 1995.

U.S. Army, *Integrated Sustainment Maintenance Expanded, Demonstration Plan,* March 1995.

Winnefeld, James A., Preston Niblack, and Dana J. Johnson, *A League of Airmen: U.S. Air Power in the Gulf War,* Santa Monica, CA: RAND, MR-343-AF, 1994.